my revision notes

CW00684258

Edexcel A-level

BIOLOGY B

Martin Rowland

HODDER
EDUCATION
AN HACHETTE UK COMPANY

Hachette UK's policy is to use papers that are natural, renewable and recyclable products and made from wood grown in sustainable forests. The logging and manufacturing processes are expected to conform to the environmental regulations of the country of origin.

Orders: please contact Bookpoint Ltd, 130 Park Drive, Milton Park, Abingdon, Oxon OX14 4SB. Telephone: (44) 01235 827720. Fax: (44) 01235 400454. Email education@bookpoint.co.uk

Lines are open from 9 a.m. to 5 p.m., Monday to Saturday, with a 24-hour message answering service. You can also order through our website: www.hoddereducation.co.uk

ISBN: 978 1 4718 5485 9

© Martin Rowland 2016

First published in 2016 by

Hodder Education,
An Hachette UK Company
Carmelite House
50 Victoria Embankment
London EC4Y 0DZ
www.hoddereducation.co.uk

Impression number   10   9   8   7   6   5   4   3   2   1

Year   2020   2019   2018   2017   2016

All rights reserved. Apart from any use permitted under UK copyright law, no part of this publication may be reproduced or transmitted in any form or by any means, electronic or mechanical, including photocopying and recording, or held within any information storage and retrieval system, without permission in writing from the publisher or under licence from the Copyright Licensing Agency Limited. Further details of such licences (for reprographic reproduction) may be obtained from the Copyright Licensing Agency Limited, Saffron House, 6–10 Kirby Street, London EC1N 8TS.

Cover photo reproduced by permission of oxilixo/Fotolia

Typeset in Integra Software Services Pvt. Ltd., Aptara, India

Printed in Spain

A catalogue record for this title is available from the British Library.

Get the most from this book

Everyone has to decide his or her own revision strategy, but it is essential to review your work, learn it and test your understanding. These *Revision Notes* will help you to do that in a planned way, topic by topic. Use this book as the cornerstone of your revision and don't hesitate to write in it — personalise your notes and check your progress by ticking off each section as you revise.

Tick to track your progress

Use the revision planner on pages 4–5 to plan your revision, topic by topic. Tick each box when you have:

- revised and understood a topic
- tested yourself
- practised the exam questions and gone online to check your answers and complete the quick quizzes

You can also keep track of your revision by ticking off each topic heading in the book. You may find it helpful to add your own notes as you work through each topic.

Features to help you succeed

Exam tips and summaries

Expert tips are given throughout the book to help you polish your exam technique in order to maximise your chances in the exam. The summaries provide a quick-check bullet list for each topic.

Typical mistakes

The author identifies the typical mistakes candidates make and explains how you can avoid them.

Now test yourself

These short, knowledge-based questions provide the first step in testing your learning. Answers are at the back of the book.

Definitions and key words

Clear, concise definitions of essential key terms are provided where they first appear.

Key words from the specification are highlighted in bold throughout the book.

Revision activities

These activities will help you to understand each topic in an interactive way.

Exam practice

Practice exam questions are provided for each topic. Use them to consolidate your revision and practise your exam skills.

Online

Go online to check your answers to the exam questions and try out the extra quick quizzes at **www.hoddereducation.co.uk/myrevisionnotes**

My revision planner

REVISED TESTED EXAM READY

Exam practice answers and quick quizzes at www.hoddereducation.co.uk/myrevisionnotes

Countdown to my exams

6–8 weeks to go

- Start by looking at the specification — make sure you know exactly what material you need to revise and the style of the examination. Use the revision planner on pages 4–5 to familiarise yourself with the topics.
- Organise your notes, making sure you have covered everything on the specification. The revision planner will help you to group your notes into topics.
- Work out a realistic revision plan that will also allow you time for relaxation. Set aside days and times for all the subjects you need to study, and stick to your timetable.
- Set yourself sensible targets. Break your revision down into focused sessions of around 40 minutes, divided by breaks. These *Revision Notes* organise the basic facts into short, memorable sections to make revising easier.

REVISED ☐

2–6 weeks to go

- Read through the relevant sections of this book and refer to the exam tips, summaries, typical mistakes and key terms. Tick off the topics as you feel confident about them. Highlight those you find difficult and look at them again in detail.
- Test your understanding of each topic by working through the 'Now test yourself' questions in the book. Look up the answers at the back of the book.
- Make a note of any problem areas as you revise, and ask your teacher to go over these in class.
- Look at past papers. They are one of the best ways to revise and to practise your exam skills. Write or prepare planned answers to the 'Exam practice' questions in this book. Check your answers online and try out the extra quick quizzes at **www.hoddereducation.co.uk/ myrevisionnotes**
- Use the revision activities to try out different revision methods. For example, you can make notes using mind maps, spider diagrams or flash cards.
- Track your progress using the revision planner and give yourself a reward when you have achieved your target.

REVISED ☐

One week to go

- Try to fit in at least one more timed practice of an entire past paper and seek feedback from your teacher, comparing your work closely with the mark scheme.
- Check the revision planner to make sure you haven't missed any topics. Brush up on any areas of difficulty by talking them over with a friend or getting help from your teacher.
- Attend any revision classes put on by your teacher. Remember, he or she is an expert at preparing people for examinations.

REVISED ☐

The day before the examination

- Flick through these *Revision Notes* for useful reminders — for example the exam tips, summaries, typical mistakes and key terms.
- Check the time and place of your examination.
- Make sure you have everything you need — extra pens and pencils, tissues, a watch, bottled water, sweets.
- Allow some time to relax and have an early night to ensure you are fresh and alert for the examinations.

REVISED ☐

My exams

A-level Biology B Paper 1: Advanced Biochemistry, Microbiology and Genetics

Date: ....

Time: ....

Location: ....

A-level Biology B Paper 2: Advanced Physiology, Evolution and Ecology

Date: ....

Time: ....

Location: ....

A-level Biology B Paper 3: General and Practical Principles in Biology

Date: ....

Time: ....

Location: ....

1 Biological molecules

Most biologically important compounds are relatively large. Three categories of molecule — carbohydrates, proteins and nucleic acids — are **polymers**. This means that they are long chains of repeated subunits called **monomers**.

- In a **condensation reaction**, two monomers become linked together. In this reaction, a molecule of water is released.
- In a **hydrolysis reaction**, a monomer is released from a polymer. In this reaction, a molecule of water is used.

Smaller compounds, such as water, and inorganic ions are also important in biology.

> A **condensation reaction** is one in which two monomers combine with the elimination of a molecule of water.
>
> A **hydrolysis reaction** is one in which a monomer is released from a polymer by the addition of a molecule of water.

Carbohydrates

Molecules of carbohydrate contain only the elements carbon, hydrogen and oxygen, in the ratio $C_x(H_2O)_y$. Table 1.1 shows the carbohydrates with which you must be familiar for Topic 1.

Table 1.1 Carbohydrates with which you must be familiar

Type of carbohydrate	Examples with which you should be familiar
Monosaccharide — the monomer from which other carbohydrates are made	Pentoses ($C_5H_{10}O_5$), e.g. ribose and deoxyribose
	Hexoses ($C_6H_{12}O_6$), e.g. glucose, fructose and galactose
Disaccharide — two monosaccharides linked together by a covalent bond called a glycosidic bond	glucose + fructose → sucrose + water
	glucose + galactose → lactose + water
	glucose + glucose → maltose + water
Polysaccharide — a large number of monosaccharides linked together by glycosidic bonds	Starch (a fuel store in plants)
	Glycogen (a fuel store in animals)
	Cellulose (a major component of plant cell walls)

Monosaccharides

REVISED

The carbon atoms in molecules of a pentose and a hexose are usually arranged in rings. The carbon atoms in each ring are numbered. Figure 1.1 shows the simplified structural formula of two **isomers** of glucose, α-glucose and β-glucose.

> **Isomers** are molecules with the same molecular formula (e.g. $C_6H_{12}O_6$) but a different structural formula.

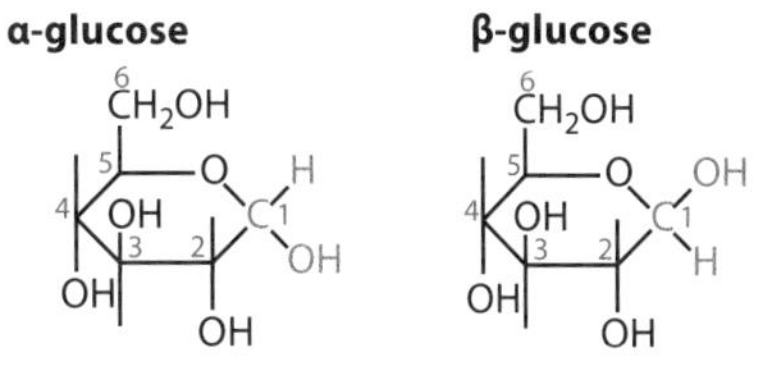

α-glucose   **β-glucose**

Figure 1.1 The structural formulae of α-glucose and β-glucose. Note that the carbon atoms are numbered from the oxygen clockwise

> **Exam tip**
>
> It **H**elps to remember the **H** is **a**bove C–1 in α-glucose but **b**elow C–1 in β-glucose.

Now test yourself

1 If the molecular formula of glucose is $C_6H_{12}O_6$, what is the molecular formula of maltose?

Answer on p. 198

Glycosidic bonds

Monosaccharides can link together during condensation reactions to form disaccharides and polysaccharides. The **covalent bond** between monosaccharides is a **glycosidic bond** (Figure 1.2). A glycosidic bond is described by the carbon atoms involved in the linkage — carbon atom 1 on one molecule with either carbon atom 4 (a 1,4 bond) or carbon atom 6 (a 1,6 bond) on the second molecule.

> A **covalent bond** is a relatively strong chemical link formed between two atoms when they share electrons.

> **Exam tip**
>
> If you are asked to draw a condensation reaction, don't forget to include the water molecule.

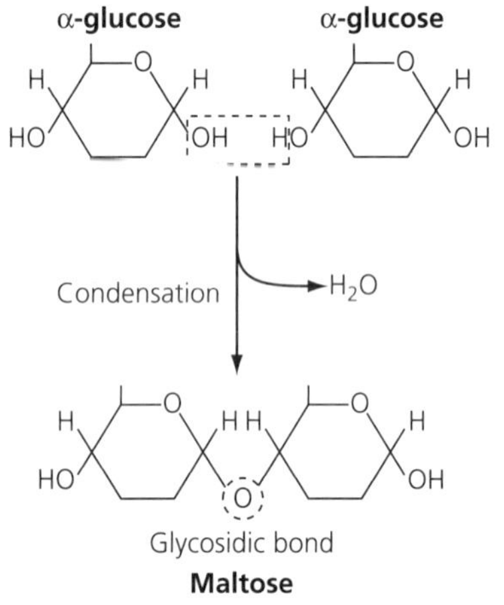

α-glucose   α-glucose

Condensation → H_2O

Glycosidic bond
Maltose

Figure 1.2 The formation of an α-1,4 glycosidic bond

Polysaccharides

REVISED

Amylose and amylopectin are polysaccharides formed from α-glucose monomers.

A molecule of **amylose** is an unbranched chain of α-glucose monomers held together only by α-1,4 glycosidic bonds (like the upper part only of Figure 1.3).

A molecule of **amylopectin** is a chain of α-glucose monomers held together by α-1,4 glycosidic bonds, but it has branches along its length caused by α-1,6 glycosidic bonds (as shown in Figure 1.3).

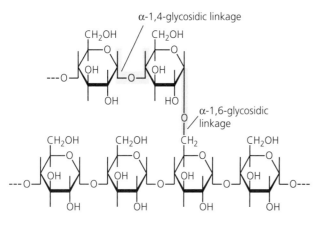

Figure 1.3 Amylopectin is a branched molecule

Starch

Starch is found in plant cells. It is a mixture of molecules of amylose and amylopectin. It is ideal as a storage molecule because:
- during hydrolysis, it releases glucose that can be used in cell respiration
- its molecules are compact because they form helices, held in shape by hydrogen bonds
- its molecules are insoluble, so do not affect the water potential of the storage cell's cytoplasm

> **Typical mistake**
>
> Students often write that respiration produces energy. This is wrong and will not be credited by examiners. An acceptable description is that respiration produces ATP.

Now test yourself

TESTED

2 Why is it important that starch does not affect the water potential of a storage cell's cytoplasm?

Answer on p. 198

Glycogen

Glycogen is found in animals cells. It is formed of highly branched molecules of amylopectin. It is an ideal storage molecule because:
- during hydrolysis, it releases glucose that can be used in cell respiration
- as it is highly branched, it can be hydrolysed faster than starch
- its molecules are insoluble, so do not affect the water potential of the storage cell's cytoplasm

> **Typical mistake**
>
> Students often confuse the spellings of similar words. Make sure you use the terms glucose, glycogen and glycosidic correctly.

Now test yourself

TESTED

3 Why does being highly branched enable glycogen to be hydrolysed faster than starch?

Answer on p. 198

Cellulose

Cellulose is a polysaccharide formed from β-glucose monomers. Cellulose is found in plant cell walls. Each cellulose molecule is a polymer of 2000 to 3000 monomers of β-glucose, held together by β-1,4 glycosidic bonds (Figure 1.4). About 200 straight chains of cellulose molecules become bound together by hydrogen bonds to form fibres. Cellulose fibres are arranged in different directions in plant cell walls, so they make the cell wall strong but flexible.

> **Revision activity**
>
> Draw a table to summarise the differences between the structures of an α-1,4 glycosidic bond, an α-1,6 glycosidic bond and a β-1,4 glycosidic bond and to show the type(s) of carbohydrates in which each is found.

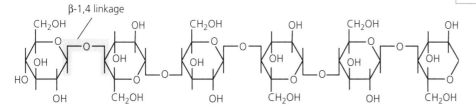

Figure 1.4 Cellulose is a polymer of β-glucose residues

Lipids

Like carbohydrates, lipid molecules contain only the elements carbon, hydrogen and oxygen. For this topic, you need to know about the structure of two lipids: triglycerides and phospholipids.

Triglycerides (fats and oils) REVISED

Figure 1.5 shows how triglycerides are formed by the condensation of one molecule of **glycerol** and three molecules of **fatty acid**. Each bond between a fatty acid and a glycerol molecule is called an **ester bond**. Triglycerides are **hydrophobic**, so can be used in waterproofing. They can be stored and used as a respiratory substrate. Fatty acids have far fewer oxygen molecules than molecules of carbohydrate, so they form a more compact fuel than carbohydrates. Triglycerides slow, or prevent, the passage of heat and ions, so are useful in insulating against heat loss or against ion leakage.

> **Hydrophobic** molecules are not wetted by, and will not interact with, water.

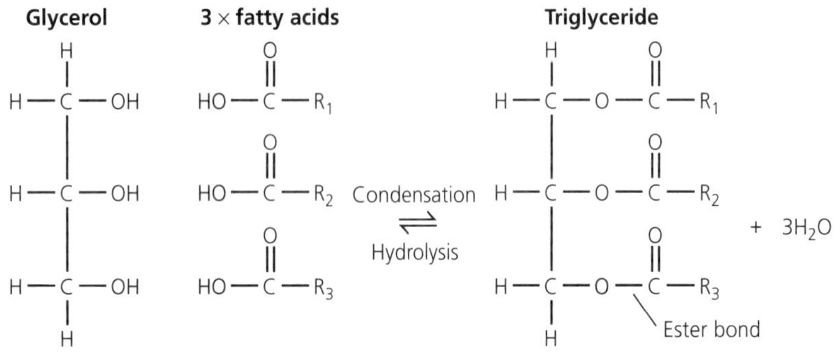

Figure 1.5 A triglyceride contains ester bonds

Saturated and unsaturated fatty acids

In a **saturated fatty acid** molecule, all carbon atoms are all held together by single covalent bonds (represented as C–C). In an **unsaturated fatty acid** molecule, at least some of the carbon atoms are held together by double covalent bonds (represented as C=C).

> **Unsaturated fatty acids** have at least one double bond (C=C) between the carbon atoms in their chain. In a saturated fatty acid, all the C–C bonds are single.

Now test yourself TESTED

4 Explain why a triglyceride is not classified as a polymer.

Answer on p. 198

Phospholipids REVISED

Phospholipids have a similar structure to triglycerides except that a phosphate group (PO_4^{3-}) replaces the fatty acid at one end of the glycerol molecule. Unlike the rest of the components of the phospholipid, the phosphate group is **hydrophilic**. When mixed with water or aqueous solutions, phospholipids form a bilayer, with the hydrophobic fatty acid 'tails' away from the water and the hydrophilic phosphate 'heads' towards the water. The basic structure of all cell membranes is a phospholipid bilayer.

> **Hydrophilic** molecules, or hydrophilic parts of molecules, will interact with, or dissolve in, water.

Proteins

In addition to carbon, hydrogen and oxygen, proteins contain the element nitrogen and some also contain sulfur. Every protein is a polymer of amino acids.

Amino acids

REVISED

Each amino acid has an amino group (NH_2) and a carboxyl group (COOH) attached by covalent bonds to a central carbon atom. There are 20 common amino acids in naturally occurring proteins. The only difference between them is the nature of the R group, shown in Figure 1.6. The simplest R group is a hydrogen atom in the amino acid glycine.

In an acidic solution, an amino group can take up a hydrogen ion to become positively charged:

$$—NH_2 + H^+ \rightarrow —NH_3^+$$

In an alkaline solution, a hydroxyl group can release a hydrogen atom to become negatively charged:

$$—COOH \rightarrow COO^- + H^+$$

Amino acids can exist as **zwitterions**, i.e.

$$NH_3^+—CR—COO^-$$

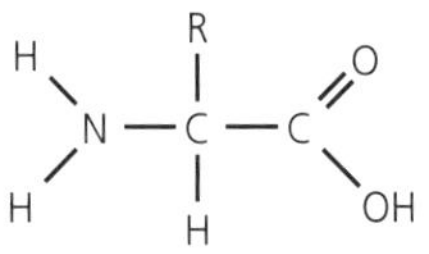

> A **zwitterion** is a neutral molecule with a positive and a negative electrical charge, though multiple positive and negative charges can be present.

Figure 1.6 The general structure of an amino acid

> **Exam tip**
>
> You will not be asked to recall the structure of the R group of any amino acid.

Peptide bonds

REVISED

When two amino acids undergo a condensation reaction, they form a **dipeptide** (Figure 1.7). The covalent bond formed between the two amino acids is a **peptide bond**. A molecule of water is released during the reaction.

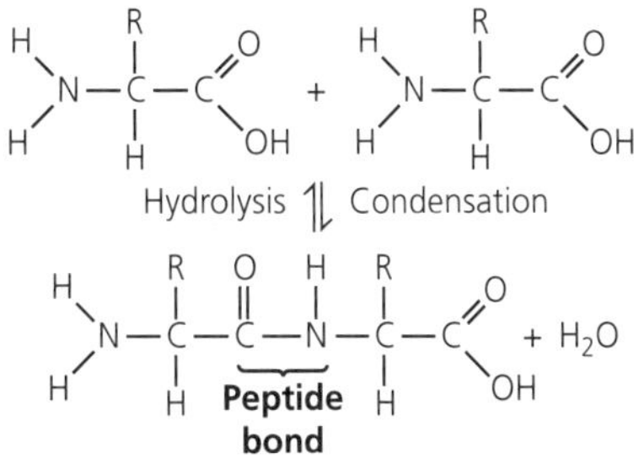

> A **dipeptide** is formed when two amino acids are joined together by a peptide bond.

Figure 1.7 The formation of a peptide bond

> **Revision activity**
>
> Use the information above to show how two molecules of glycine produce a dipeptide. Use structural formulae in your diagram.

A **polypeptide** is a polymer containing a large number of amino acids held together by peptide bonds. The number and sequence of amino acid acids makes one polypeptide different from another. The sequence of amino acids in each polypeptide is controlled by a specific gene.

Levels of organisation in protein structure

REVISED

There are four levels of organisation in protein structure:
- **Primary structure** — the sequence of amino acids in a polypeptide.
- **Secondary sequence** — the way in which sections of a polypeptide chain fold into an α-helix (shaped like a coiled spring) or a β-pleated sheet (shaped like a folded sheet of paper), held together by hydrogen bonds between the hydrogen of one amino group and the oxygen of an adjacent carboxyl group.
- **Tertiary structure** — the way in which a polypeptide folds into a complex, three-dimensional globular shape, held together by hydrogen bonds, disulfide bonds between cysteine residues and ionic bonds between charged *R* groups.
- **Quaternary structure** — two or more polypeptides combine together to form a functional protein. Not all proteins have a quaternary structure.

Fibrous and globular proteins

REVISED

Fibrous proteins are long and straight, whereas globular proteins are spherical.

Collagen

Collagen is a fibrous protein. It is a component of bone and tendons. It provides strength to these tissues because:
- its molecule is a helix made of three polypeptide chains (Figure 1.8)
- the three polypeptides are held together by a large number of covalent bonds and hydrogen bonds

Haemoglobin

Haemoglobin is a globular protein. It accounts for about 95% of the dry mass of our red blood cells. It is able to transport oxygen because:
- its molecule has four polypeptide chains (Figure 1.8)
- each polypeptide chain has an iron-containing haem group
- each haem group combines with an oxygen molecule in the lungs and releases this oxygen molecule in the body tissues

Collagen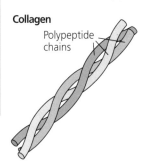

Haemoglobin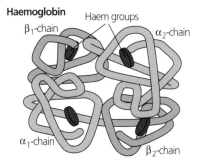

Figure 1.8 Collagen is a fibrous protein and haemoglobin is a globular protein

Now test yourself

TESTED

5 How many levels of protein organisation can you see in the haemoglobin molecule in Figure 1.8?

Answer on p. 198

DNA and protein synthesis

Two types of nucleic acid are involved in the production of proteins:

- deoxyribonucleic acid (**DNA**)
- ribonucleic acid (**RNA**)

Both are polymers of nucleotides.

Nucleotides

REVISED

A **nucleotide** is the monomer from which all nucleic acids are made. Each nucleotide consists of:

- a **pentose** — ribose in RNA or deoxyribose in DNA (Figure 1.9)
- a nitrogen-containing **organic base**, which can be a single-ringed pyrimidine (cytosine or thymine in DNA; cytosine or uracil in RNA), or a double-ringed purine (adenine or guanine in both DNA and RNA)
- a **phosphate group**

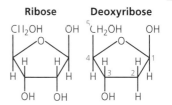

Figure 1.9 Ribose and deoxyribose

The structure of DNA

REVISED

A DNA molecule consists of two polynucleotide chains (Figure 1.10). Each polymer consists of nucleotides held together by **phosphodiester bonds** between carbon-3 of one deoxyribose and carbon-5 of the next deoxyribose in the chain. The two strands are **antiparallel**, i.e. they run in opposite directions, as shown in Figure 1.10. The two strands are held together by hydrogen bonds between specific pairs of organic bases:

- adenine pairs with thymine (A–T)
- cytosine pairs with guanine (C–G)

The two strands are twisted into a helix: the so-called **double-helix** model.

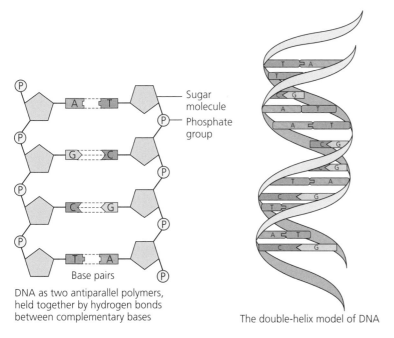

DNA as two antiparallel polymers, held together by hydrogen bonds between complementary bases

The double-helix model of DNA

Figure 1.10 A molecule of DNA consists of two antiparallel polynucleotides twisted into a helix

Now test yourself

6 Use information from Figure 1.9 to describe the difference between ribose and deoxyribose.
7 Use information from page 13 to define the term 'phosphodiester bond'.

Answers on p. 198

The semi-conservative replication of DNA

Before a cell divides, it replicates its DNA (Figure 1.11). This ensures that the new cells are **clones** of the original cell.

> **Clones** are cells or organisms that have identical DNA to other cells or organisms.

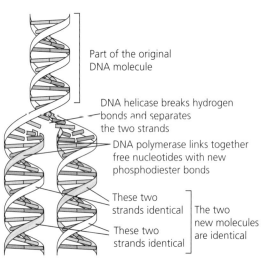

Part of the original DNA molecule

DNA helicase breaks hydrogen bonds and separates the two strands

DNA polymerase links together free nucleotides with new phosphodiester bonds

These two strands identical

These two strands identical

The two new molecules are identical

Figure 1.11 **The semi-conservative replication of DNA**

During the semi-conservative replication of DNA, **DNA helicase** breaks the hydrogen bonds holding the two polynucleotides together and separates the two strands. Each DNA strand acts as a template for the production of a new strand. The unpaired bases on each template strand attract free nucleotides with bases that are complementary to themselves. Hydrogen bonds form between the new base pairs. **DNA polymerase** catalyses the formation of new phosphodiester bonds between the newly added nucleotides. It catalyses the formation of each phosphodiester bond between the hydroxyl group on carbon-3 of the deoxyribose at the end of the developing polynucleotide and the phosphate group on carbon-5 of the deoxyribose of the nucleotide being added. Where one DNA strand is replicated in short fragments, **DNA ligase** catalyses the linking of these fragments to make a continuous chain.

Now test yourself

8 DNA polymerase can form a phosphodiester bond only between the hydroxyl group on carbon-3 of the deoxyribose at the end of the developing polynucleotide and the phosphate group on carbon-5 of the deoxyribose of the nucleotide being added.
   (a) Use your existing knowledge to suggest why.
   (b) Suggest what effect this will have on the development of the new nucleotide chains during DNA replication.

Answers on p. 198

Typical mistake

Students often tell examiners that DNA polymerase catalyses the formation of DNA base pairs. It does not; instead, it catalyses the formation of phosphodiester bonds between nucleotides that have already been added to a developing polynucleotide chain.

mRNA and tRNA

REVISED

Unlike DNA, an RNA molecule has only one strand of nucleotides:
- A messenger RNA (**mRNA**) molecule is a straight polynucleotide chain containing a few hundred to a few thousand RNA nucleotides.
- Transfer RNA (**tRNA**) contains only about 75 RNA nucleotides. Figure 1.12 shows:
  - it is folded into the shape of a clover leaf, held in place by hydrogen bonds between a few complementary base pairs (adenine with uracil and cytosine with guanine)
  - the end of one of the chains has a region that attaches to a specific amino acid
  - the middle loop has a region of three bases, called an **anticodon**

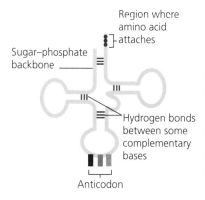

Figure 1.12 **A molecule of tRNA**

The genetic code

REVISED

The sequence of organic bases in a molecule of DNA contains the code for making polypeptides. Each amino acid is encoded by a sequence of three bases, i.e. a **base triplet**. All 64 possible triplets of DNA bases and the amino acids they encode are shown in Table 1.2. This is the **genetic code** and it is:
- **universal** — in all organisms, the same base triplet codes for the same amino acid
- **non-overlapping** — each base forms part of only one triplet. For example, a sequence of CGTATC will always be 'read' as two base triplets — CGT and ATC — and **not** as four triplets CGT, GTA, TAT and ATC
- **degenerate** — some amino acids are encoded by more than one base triplet

In addition, some of the codons act as 'reading instructions' for ribosomes. ATG, the code for methionine, is also a 'start' code, and TAA, TAG and TGA are 'stop' codes.

Exam tip

You don't have to remember any of the base triplet and amino acid combinations in Table 1.2. If needed, they will always be provided in the exam question.

Table 1.2 **The genetic code using DNA bases on the sense strand**

First position		Second position			Third position
	T	C	A	G	
T	Phenylalanine	Serine	Tyrosine	Cysteine	T
	Phenylalanine	Serine	Tyrosine	Cysteine	C
	Leucine	Serine	(stop)	(stop)	A
	Leucine	Serine	(stop)	Tryptophan	G
C	Leucine	Proline	Histidine	Arginine	T
	Leucine	Proline	Histidine	Arginine	C
	Leucine	Proline	Glutamine	Arginine	A
	Leucine	Proline	Glutamine	Arginine	G
A	Isoleucine	Threonine	Asparagine	Serine	T
	Isoleucine	Threonine	Asparagine	Serine	C
	Isoleucine	Threonine	Lysine	Arginine	A
	Methionine	Threonine	Lysine	Arginine	G
G	Valine	Alanine	Aspartic acid	Glycine	T
	Valine	Alanine	Aspartic acid	Glycine	C
	Valine	Alanine	Glutamic acid	Glycine	A
	Valine	Alanine	Glutamic acid	Glycine	G

Now test yourself

TESTED

9 How can you tell that Table 1.2 contains DNA base triplets and not RNA base triplets?
10 In addition to encoding methionine, the DNA triplet ATG is a 'start' code for ribosomes. What is the equivalent mRNA base sequence for this 'start' code?

Answers on p. 198

Using the DNA code to make a polypeptide

REVISED

A **gene** is a region in the **sense strand** of a DNA molecule that codes for a particular polypeptide. The code for the polypeptide is determined by the sequence of organic bases within the gene.

> A **gene** is a DNA base sequence that encodes the amino acid sequence of a polypeptide.

> The **sense strand** is the strand of a DNA molecule with the base sequence that carries the genetic code for the amino acid sequence of a polypeptide.

> The **antisense strand** is the strand of a DNA molecule with a base sequence that is complementary to that of the sense strand. It is this strand that is used as a template for the base sequence of mRNA.

The sequence of events by which the DNA code is used to make a polypeptide is summarised in Figure 1.13. The DNA base sequence of the **antisense strand** is used as a template for the base sequence of a molecule of messenger RNA (mRNA) in a process called **transcription**. The mRNA travels from the nucleus through pores in the nuclear envelope to the ribosomes. Free amino acids are carried from the cytoplasm to the ribosomes by molecules of transfer RNA (tRNA). Ribosomes 'read' the mRNA code and assemble the amino acids into a polypeptide in a process called **translation**.

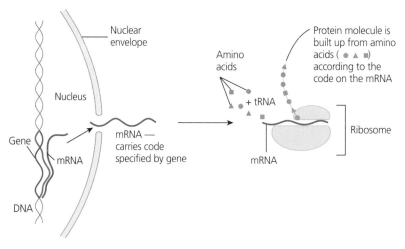

Figure 1.13 A summary of polypeptide synthesis in a eukaryotic cell

Exam tip

If you remember that the alphabetical order of the first letter after 'trans' is also the order of the two processes — trans**c**ription and trans**l**ation — you will not confuse the two in an exam.

Sense and antisense strands

DNA is a two-stranded molecule. The base sequence of the sense strand carries the code for protein synthesis. The other strand is the antisense strand and its base sequence is complementary to that of the sense strand.

By transcribing the antisense strand, mRNA has a base sequence that is *complementary* to the antisense strand of the DNA and, therefore, is the *same* as the sense strand (except that uracil replaces thymine).

Exam tip

Students often confuse the terms 'sense strand' and 'antisense strand'. As long as you make it clear that only one of the two DNA strands is transcribed, the examiner will reward you — even if you use neither term.

Transcription

Transcription (Figure 1.14) takes place as follows. The enzyme **RNA polymerase** attaches to the start of the gene. The DNA 'unwinds' as the hydrogen bonds between complementary base pairs are broken. Free RNA nucleotides pair with complementary nucleotides on the exposed antisense strand of DNA. The RNA polymerase moves along the gene, catalysing the formation of phosphodiester bonds between the RNA nucleotides, forming a molecule of mRNA. When complete, the mRNA leaves the nucleus via a pore in the nuclear envelope and attaches to a ribosome in the cytoplasm.

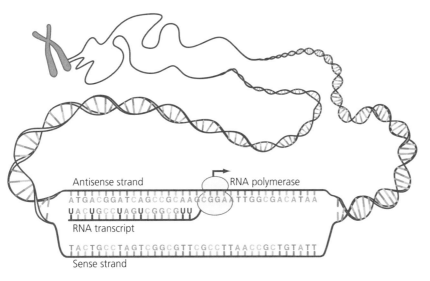

Figure 1.14 During transcription, the base sequence of the antisense strand becomes the 'sense' base sequence of mRNA

Typical mistake

Students often write that RNA polymerase adds complementary bases to the developing mRNA molecule. Be sure to use the correct terminology: it adds complementary RNA nucleotides.

Introns and exons

In eukaryotic cells, genes contain sections of non-coding DNA, called **introns**, between the coding sections, called **exons**. During transcription, the entire gene is transcribed to form pre-mRNA. Before it leaves the nucleus, the introns are removed from the pre-mRNA to form mature RNA. During editing, different combinations of exons can be assembled from transcription of one gene. In this way, a single gene can encode more than one polypeptide. Figure 1.15 summarises this process.

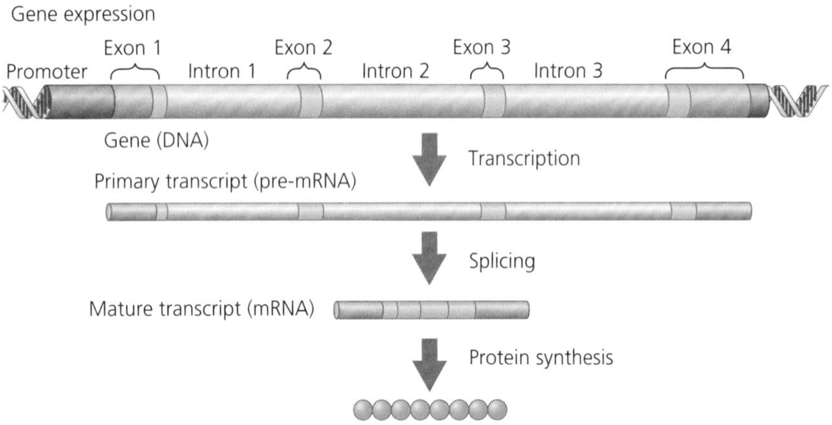

Gene expression

Figure 1.15 In eukaryotic cells, mRNA is edited before it can be translated

Exam tip

Remember that it is the **ex**ons that are **ex**pressed, i.e. are used to encode amino acid sequences.

Now test yourself

TESTED

11 What is the difference between an exon and an intron?
12 Suggest why mRNA in prokaryotic cells is not edited before it is used by the ribosomes.

Answers on p. 198

The mRNA now carries a base sequence that is the same as that of the sense strand of the gene. The base triplets on this mRNA molecule are referred to as **codons**.

Codons are base triplets on **mRNA** that encode an amino acid.

Translation

During translation, amino acids are joined together by ribosomes in a sequence that is determined by the base sequence of the mRNA.

1 Ribosomes have two binding sites. The first binding site binds to the beginning of the mRNA molecule. The first codon on the mRNA is always AUG (the 'start' codon).

2 A molecule of tRNA with the complementary anticodon, UAC, arrives carrying its specific amino acid (Figure 1.16).

3 In Figure 1.17, the ribosome has moved along the molecule of mRNA and now has both binding sites filled by tRNA molecules. The codon now held in the second binding site is CAC. A tRNA molecule with the complementary anticodon, GUG, has arrived carrying its specific amino acid.

4 Accompanied by the hydrolysis of ATP, enzymes in the ribosome catalyse the formation of a peptide bond between the two amino acids.

5 The ribosome continues to move along the mRNA one codon at a time and the peptide grows one amino acid at a time (Figure 1.17).

6 The final codon is a 'stop' codon. Once reached, the ribosome detaches from the mRNA and releases the polypeptide.

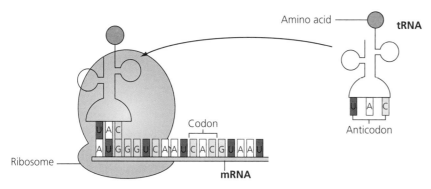

Figure 1.16 Translation, steps 1 and 2

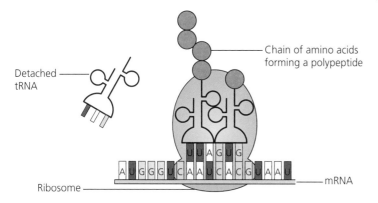

Figure 1.17 Translation, steps 3 to 6

Gene mutations

A gene mutation results if an error occurs during DNA replication that changes the DNA base sequence. Such errors occur randomly at a rate of about 1 in 10^7 base additions and might affect the encoded polypeptide such that it is no longer functional. Many do not, however, because:

- they occur in non-coding regions within genes (i.e. within introns)
- they occur in non-coding regions between genes
- the new base triplet still encodes the same amino acid (remember the code is degenerate)
- the new amino acid sequence has no effect on the key parts of the tertiary structure of the encoded polypeptide, so it can still function normally

Types of gene mutation

Gene mutations can occur in different ways:

- **base deletion** — one nucleotide is lost from the sequence
- **base insertion** — a new nucleotide is inserted into the sequence
- **base substitution** — one nucleotide is used in place of another

Table 1.3 shows the effects of these gene mutations. Note:

- A deletion or insertion mutation affects all the DNA triplets, and the encoded amino acid sequence, following the mutation. This is called a frameshift.
- A substitution mutation affects only the one base triplet and the amino acid it encodes. This is called a **point mutation**.

> A **frameshift** is a genetic mutation caused by a deletion or insertion in a DNA sequence that shifts the way the sequence is read.

Table 1.3 **The effects of gene mutation**

Base sequence on sense DNA strand	AGA	TAC	GCA	CAC	ATG	CGC
Encoded amino acid sequence	Arginine	Tyrosine	Alanine	Histidine	Methionine	Arginine
Base sequence after base deletion	AGT	ACG	CAC	ACA	TGC	GC?
Encoded amino acid sequence	Serine	Threonine	Histidine	Threonine	Cysteine	Alanine
Base sequence after base insertion	AGC	ATA	CGC	ACA	CAT	GCG
Encoded amino acid sequence	Serine	Isoleucine	Arginine	Threonine	Histidine	Alanine
Base sequence after base substitution	AGT	TAC	GCA	CAC	ATG	CGC
Encoded amino acid sequence	Serine	Tyrosine	Alanine	Histidine	Methionine	Arginine

Now test yourself

13 The sense strand of a DNA molecule has the base sequence CTAGCC. Give the anticodons of the tRNA molecules that will carry amino acids to the mRNA produced from this DNA.

14 In which DNA base triplet would a point mutation always result in a change in the encoded amino acid? Explain your answer.

Answers on p. 198

Point mutation and sickle-cell anaemia

Haemoglobin is a protein with two α-globulins and two β-globulins (look back to Figure 1.8). A particular point mutation in the gene encoding the β-globulins results in:

● a base triplet becoming GTG instead of GAG
● the incorporation into the β-globulins of the amino acid valine instead of the amino acid glutamic acid

This sounds like a tiny change, but the effect is striking. Red blood cells containing haemoglobin affected by the mutation tend to collapse, becoming crescent or sickle-shaped. In this condition, the red blood cells do not carry oxygen effectively.

The mutated gene is inherited. If all of a person's haemoglobin is affected, the condition is lethal. If only half of a person's haemoglobin is affected, they suffer anaemia but have increased resistance to malaria. As a result of natural selection, the condition is common in people originating from areas where malaria is endemic.

Enzymes

Enzymes are globular proteins that catalyse a wide range of intracellular and extracellular reactions. We can represent the action of an enzyme in a simple equation:

substrate(s) + enzyme → enzyme–substrate complex → product(s) + enzyme

Like all catalysts, enzymes:

● speed up the reaction they catalyse by reducing the **activation energy** for that reaction
● are needed in much smaller concentrations than the reactants
● are not chemically changed by the reaction and can be recovered at the end

Activation energy is the energy 'barrier' that must be overcome to enable a reaction to proceed.

The induced-fit hypothesis

REVISED

Unlike most inorganic catalysts, enzymes are specific because they are globular proteins. Part of each enzyme is the **active site** to which the reactants (the substrates) bind when forming an enzyme–substrate complex. A substrate can only combine with an active site that has a tertiary structure that is complementary its own.

Typical mistake

Students often state that the active site of an enzyme and the substrate molecule have the same shape. Make sure you always make it clear that the shapes are complementary.

Typical mistake

Students often state, or infer, that the active site is on the substrate. Ensure you always make it clear that the active site is part of the enzyme molecule.

Figure 1.18 shows a model of enzyme action, called the **induced-fit hypothesis**. The tertiary structure of the enzyme's active site is not normally an exact complement to the shape of its substrate. Binding of the substrate molecule(s) changes the tertiary structure of the active site, causing it to become an exact complementary fit.

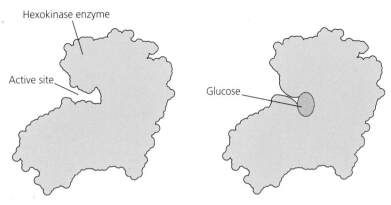

Figure 1.18 **The induced-fit model of enzyme action**

Answer on p. 198

Now test yourself

TESTED

15 Maltase is an enzyme that hydrolyses the disaccharide maltose into glucose. It will not hydrolyse other disaccharides, such as sucrose or lactose. Explain why.

The initial rate of an enzyme-catalysed reaction

REVISED

The rate of an enzyme–catalysed reaction is the change in the concentration of substrate or of products in a given period of time.

Figure 1.19 shows how the rate of reaction slows as the substrate molecules are used up. This means that knowing the value of the rate of reaction loses its meaning over a long period of time. Consequently, the initial rate of reaction is a more useful measure of the properties of an enzyme. We can find the initial rate of reaction by drawing a tangent to the start of the curve, like the one in Figure 1.19, and finding its gradient.

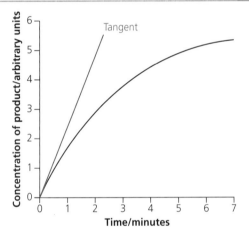

Figure 1.19 **With a fixed concentration of substrate, the rate of an enzyme-catalysed reaction slows**

Now test yourself

TESTED

16 The curve in Figure 1.19 levels off after 7 minutes. Explain why.
17 Calculate the initial rate of the enzyme-catalysed reaction in Figure 1.19.
18 Sketch a new curve on Figure 1.19 to show the concentration of substrate.

Answers on p. 198

Factors affecting the rate of enzyme activity

REVISED

The rate of enzyme activity depends on:
- the rate of collisions between molecules of the enzyme and those of its substrate — these collisions must be at the correct alignment and speed to enable an enzyme–substrate complex to form
- the ability of the enzyme to maintain the appropriate tertiary structure of its active site

Temperature

An increase in temperature affects an enzyme-catalysed reaction in two ways:
- It increases the rate of collisions between enzyme and substrate molecules.
- It **denatures** the enzyme molecules, changing the tertiary structure of their active sites.

These effects are summarised in Figure 1.20, which shows that, where the two balance, the enzyme operates at its maximum rate. This is the **optimum temperature** for that enzyme.

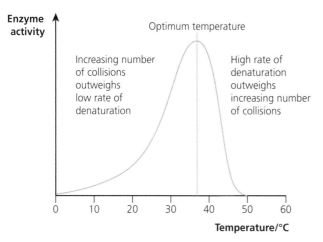

Figure 1.20 Enzyme activity is affected by temperature

pH

Amino acids can ionise in different ways, depending on the **pH** of the solution. Such changes alter the ionic bonds within the globular protein forming the enzyme and result in changes in its tertiary structure, i.e. they denature the molecule. Only at the optimum pH is the tertiary structure of the enzyme's active site complementary to that of its substrate (Figure 1.21).

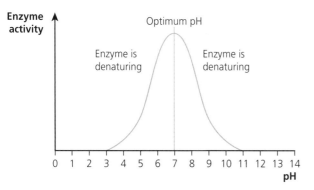

Figure 1.21 Enzyme activity is affected by pH

> **Exam tip**
>
> All molecules have a three-dimensional shape, so examiners expect you to use the term 'tertiary structure' when referring to the active site of an enzyme.

> **pH** is a measure of acidity or alkalinity, expressed as '$-\log_{10}$ hydrogen ion concentration'. The values run from 1 (highly acidic), through 7 (neutral) to 14 (highly alkaline).

Substrate concentration

At low substrate concentrations, there are fewer enzyme–substrate collisions and the rate of reaction is slow. The substrate concentration is said to be a limiting factor.

Enzyme concentration

At low enzyme concentrations, there are fewer enzyme–substrate collisions and the rate of reaction is slow. The enzyme concentration is said to be a limiting factor.

The presence of inhibitors

An **inhibitor** is a substance that combines with an enzyme molecule and reduces the rate of the reaction it catalyses.

Competitive inhibitors have a similar shape to that of the true substrate. They can bind to the active site of the enzyme, temporarily preventing its ability to form an enzyme–substrate complex. Increasing the concentration of substrate reduces the effect of a competitive inhibitor as it makes a substrate–enzyme collision more likely than an inhibitor–enzyme collision.

Non-competitive inhibitors do not have a similar shape to the true substrate. They bind to the enzyme at an **allosteric site**, i.e. a site other than the active site. In doing so, they change the tertiary structure of the enzyme molecule, so that its active site is no longer complementary to the substrate.

> **Revision activity**
>
> Draw a table to summarise how environmental factors affect the ability of enzymes to catalyse their reaction.

Inorganic ions

An ion is an atom or a group of chemically linked atoms that has lost or gained one or more electrons. If it has gained one or more electrons, the ion has a negative charge and is called an **anion**. If it has lost one or more electrons, the ion has a positive charge and is called a **cation**.

The following are examples of inorganic ions that are important in plant biology:
- Calcium ions (Ca^{2+}) are used to form calcium pectate in plant cell walls.
- Magnesium ions (Mg^{2+}) are used to produce chlorophyll.
- Phosphate ions (PO_4^{3-}) are used to make ADP and ATP.
- Nitrate ions (NO_3^-) are used to make amino acids and nucleic acids.

Now test yourself

19 Is a phosphate ion an anion or a cation?

Answer on p. 198

> **Revision activity**
>
> Sketch two graphs, one to show the effect of enzyme concentration on the rate of reaction and the second to show the effect of substrate concentration on the rate of reaction. Label each with explanations of any change in the rates.

> **Typical mistake**
>
> Students often confuse the reason why a curve representing an enzyme-catalysed reaction levels off. If the y-axis shows the rate of reaction, the reaction has not stopped when the curve is level — it is proceeding at a constant rate.

> An **allosteric site** is a part of an enzyme molecule other than its active site.

> **Exam tip**
>
> Graphs relating to enzyme activity often have curves with similar shapes. Read the labels carefully on both axes before answering the question.

TESTED

Water

A molecule of water (H_2O) contains two hydrogen atoms bound to one oxygen atom by covalent bonds.

The dipole nature of water

REVISED

Although a molecule of water has no overall charge, it has a **dipole** nature because the larger nucleus of the oxygen atom draws electrons away from the hydrogen atoms, resulting in:

- a small positive charge (δ^+) on each hydrogen atom (each hydrogen becomes a positive pole)
- a small negative charge (δ^-) on the oxygen atom (oxygen becomes a negative pole)

As a result, **hydrogen bonds** can form between water molecules (Figure 1.22).

> A **hydrogen bond** is a relatively weak chemical link formed by the attraction between a weakly positive atom and a weakly negative atom.

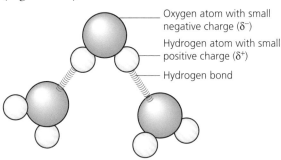

Oxygen atom with small negative charge (δ^-)

Hydrogen atom with small positive charge (δ^+)

Hydrogen bond

Figure 1.22 Hydrogen bonds cause the cohesion of water molecules

The properties of water

REVISED

The hydrogen bonding of water molecules gives water a number of properties that are important in biology:

- **Cohesion** — molecules of water 'stick together'. This is important in, for example, the movement of water in plants.
- **High specific heat capacity** — a large amount of heat is required to break the hydrogen bonds holding water molecules together. Seas and lakes therefore tend to have temperatures that are more stable than air temperature.
- **High latent heat of vaporisation** — a large amount of heat is required to turn liquid water into water vapour. The evaporation of water from the surface of organisms (e.g. water from leaves or sweat on the skin of a human) has a strong cooling effect.
- **Polar solvent** — because water molecules are polar, it is an important solvent in which ions and charged compounds can all dissolve. The reactions inside cells all occur in aqueous solution.
- **High surface tension** — at a water–air interface, the cohesion between water molecules gives the water surface high surface tension.
- **Incompressibility** — water cannot be easily compressed, making it ideal for supporting plant tissues and the bodies of soft-bodied animals.
- **Maximum density at 4°C** — even though a lake or pond might be covered in ice, there will be a layer of warmer water beneath it.

Exam practice

1 The diagram represents a molecule of the amino acid glycine.
   (a) Draw a diagram to show a condensation reaction
       between two glycine molecules. [2]
   (b) What name is given to the bond formed between the
       two amino acids? [1]
   (c) Explain why changes in pH affect the tertiary structure of a polypeptide. [3]

> **Exam tip**
>
> When a question uses the command word 'explain', make sure you
> give reasons in your answer.

2 Amylase is an enzyme that catalyses the hydrolysis of glycogen and starch into maltose.
   (a) What type of molecule is maltose? [1]
   (b) Explain how an enzyme speeds the rate of the reaction it catalyses. [2]
   (c) Amylase can hydrolyse glycogen and starch but not cellulose. Explain why amylase is
       unable to hydrolyse cellulose. [3]

3 The diagram represents a phospholipid.
   (a) What is represented by the parts of the
       molecule labelled **A**, **B** and **C**? [3]
   (b) Explain how the properties of phospholipids
       determine the structure of cell surface
       membranes. [3]

4 A student investigated the effect of substrate concentration on the initial rate of an enzyme-
   catalysed reaction. During her investigation, she measured the concentration of products.
   (a) Give two variables that the student should have controlled in her investigation and explain
       why each should be controlled. [2]
   (b) Describe how the student could find the initial rate of reaction at each substrate
       concentration. [3]

> **Exam tip**
>
> When a question uses the command word 'give', you need write no
> more than the name required by the question.

The diagram shows a sketch graph of her processed data.

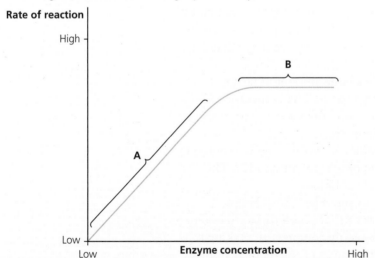

(c) Explain the value of sketching a graph of processed data. [1]
(d) Explain the shape of the curve over region **A**. [2]
(e) Explain the shape of the curve over region **B**. [2]

➡

5 The diagram represents DNA replication.

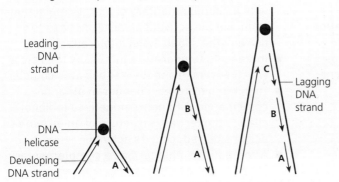

(a) What is the evidence in the diagram that DNA replication is semi-conservative? [2]
(b) What determines the sequence of nucleotides on each developing strand? [3]
(c) The arrows show the direction in which new nucleotides are added to the developing strands. Explain why they point in different directions. [3]
(d) Describe how fragments **A**, **B** and **C** are joined together. [2]

> **Exam tip**
>
> When a question includes the words 'what is the evidence in the diagram', you will not gain marks unless you relate your answers to evidence in the diagram.

6 Cosmologists are searching the universe for planets on which conditions might be suitable for life. The presence of water is an important feature in their search. Explain why water is so important to life. [6]

Answers and quick quiz 1 online

ONLINE

Summary

Carbohydrates, lipids and proteins

- Monosaccharides are important respiratory substrates. Polysaccharides are polymers of monosaccharides chemically linked by glycosidic bonds.
- Glycogen and starch are polysaccharides that are important fuel stores. The polysaccharide cellulose adds flexibility and strength to plant cell walls.
- Triglycerides are a type of lipid comprising three fatty acids molecules chemically linked by ester bonds to a molecule of glycerol. Triglycerides can act as a fuel store; their hydrophobic nature also relates to their roles in waterproofing and insulation.
- Proteins are polymers of amino acids chemically linked by peptide bonds. Collagen is an example of a fibrous protein. Enzymes and haemoglobin are examples of globular proteins.

DNA and protein synthesis

- Nucleic acids are polymers of nucleotides chemically linked by phosphodiester bonds. An RNA nucleotide contains the pentose ribose and one of the organic bases adenine, cytosine, guanine or uracil. A DNA nucleotide contains the pentose deoxyribose and one of the organic bases adenine, cytosine, guanine or thymine.
- DNA molecules are double stranded and extremely long. Their base sequence encodes the amino acid sequence of a large number of polypeptides. Messenger RNA (mRNA) molecules are single stranded and relatively short. Their base sequence encodes the amino acid sequence of a single polypeptide. Transfer RNA (tRNA) molecules are about 75 bases long; each carries a specific amino acid to the ribosomes during polypeptide formation.

- During the semi-conservative replication of DNA, the two strands of each molecule separate and act as a template for the formation of a new complementary strand. The stands are normally held together by hydrogen bonds between complementary base pairs — adenine with thymine and cytosine with guanine.
- Polypeptide synthesis is the result of two processes. During transcription, the base sequence of the antisense strand of DNA of one gene acts as a template for the formation of a molecule of mRNA. During translation, the base sequence of a molecule of mRNA is used to assemble amino acids in the correct order.
- Gene mutations occur during DNA replication when a DNA base is added, deleted or substituted. Additions and deletions cause frameshifts; a substitution causes a point mutation. Sickle-cell anaemia is an example of a point mutation.

Enzymes

- Enzymes act as catalysts in intracellular and extracellular reactions.
- Part of the tertiary structure of an enzyme molecule is its active site. Following appropriate collisions, this active site binds with its complementary substrate molecule(s), reducing the activation energy needed to convert the substrate(s) into product(s).
- Increasing temperature increases the rate at which enzymes and their substrates collide.
- High temperatures, changes in pH and combination with non-competitive inhibitors change the tertiary structure of an enzyme's active site, reducing its activity.
- A competitive inhibitor has a complementary shape to an enzyme's active site and, by binding to it, temporarily blocks enzyme–substrate collisions. A non-competitive inhibitor binds to an enzyme at an allosteric site, changing the tertiary structure of the enzyme.

Inorganic ions and water

- Inorganic ions are used by plants in producing organic compounds such as amino acids, chlorophyll and ATP.
- The dipole nature of water molecules gives water properties that are important in living organisms.

2 Cells, viruses and reproduction of living things

Eukaryotic and prokaryotic cell structure and function

Cell theory

REVISED

The cell theory is a unifying concept in biology. Cells are the fundamental unit of structure, function and organisation in all living organisms. In complex multicellular organisms:

- cells of a similar origin and function form a **tissue**, e.g. epithelial tissue
- tissues are organised into **organs**, e.g. the heart has epithelial tissue and muscle tissue
- tissues and organs are organised into **systems**, e.g. the reproductive system

Comparing prokaryotic and eukaryotic cells

REVISED

The cells of bacteria are prokaryotic, i.e. they lack a nucleus. The cells of protoctists, fungi, plants and animals are eukaryotic, i.e. they possess a nucleus. Table 2.1 summarises the major differences between prokaryotic and eukaryotic cells.

Table 2.1 **Comparing prokaryotic and eukaryotic cells (S represents a Svedberg unit — a measure of sedimentation rate)**

Feature	Prokaryotic cells	Eukaryotic cells
Size	Typically 1 μm to 10 μm	Typically 10 μm to 800 μm
Cell wall	A wall of peptidoglycan — a polymer of amino acids and monosaccharides — present	A wall of chitin present in fungi; a wall of cellulose present in plants and algae; absent in animals
Cell surface membrane	Present	Present
Nucleus	Absent	Present and separated from cytoplasm by a nuclear envelope
DNA molecules	The main DNA molecule is relatively short, circular and is not associated with proteins (i.e. is naked), forming a **nucleoid** In addition, a variable number of even shorter **plasmids** may be present	Within the nucleus, long, linear, DNA molecules are wound around proteins forming **chromosomes** Within chloroplasts and mitochondria, short, circular naked DNA molecules are found
Membrane-bound organelles	Absent	A variety of different membrane-bound organelles are present
Ribosomes	70S ribosomes present in cytoplasm	80S ribosomes in cytoplasm and on rough endoplasmic reticulum 70S ribosomes present in chloroplasts and mitochondria

Only bacteria have prokaryotic cells. Figure 2.1 shows the generalised **ultrastructure** of a bacterium. It also summarises the function of the cell components that you should know.

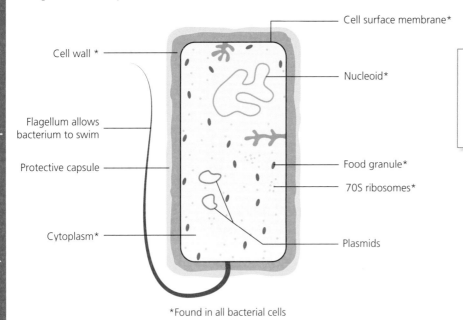

*Found in all bacterial cells

Figure 2.1 The ultrastructure of a generalised bacterial cell

> The **ultrastructure** is the features of a cell that can be seen with an electron microscope but not with a light microscope.

A prokaryotic cell has:
- an outer **cell wall** of peptidoglycan that maintains the shape of the cell
- a **cell surface membrane** that controls movement of substances into and out of the cell
- a **flagellum** that enables some bacteria to swim
- cytoplasm containing few structures, but including:
  ○ a single **nucleoid** — a circular, naked DNA molecule carrying most of the bacterial genes
  ○ **plasmids** — small, circular, naked DNA molecules containing few genes
  ○ **70S ribosomes** that synthesise bacterial polypeptides
  ○ **food granules** — fuel stores

> **Exam tip**
>
> You should not refer to the DNA of a bacterium as a chromosome as it is neither linear nor associated with proteins.

Gram-positive and Gram-negative bacteria

All bacteria have a cell wall made of **peptidoglycan** — polymers of amino acids and monosaccharides. Some have additional layers, as shown in Figure 2.2. These additional layers change the way in which bacterial cells react with a stain called crystal violet. This forms the basis of the Gram stain technique.

Gram-positive bacteria have thick walls of peptidoglycan. This wall becomes purple when stained with crystal violet. **Gram-negative bacteria** have thin walls of peptidoglycan with an additional outer layer. This outer layer prevents crystal violet staining the peptidoglycan, so these bacteria do not stain purple.

The importance of being able to identify Gram-positive and Gram-negative bacteria is that the outer lipid-rich layer of Gram-negative bacteria is relatively impermeable to antibiotics, including penicillin. Unlike Gram-positive bacteria, Gram-negative bacteria are not susceptible to these antibiotics.

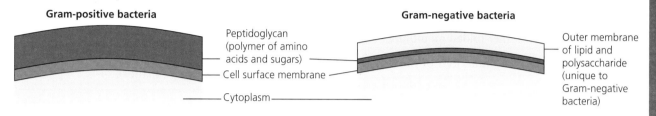

Gram-positive bacteria

Gram-negative bacteria

Peptidoglycan (polymer of amino acids and sugars)

Cell surface membrane

Cytoplasm

Outer membrane of lipid and polysaccharide (unique to Gram-negative bacteria)

Figure 2.2 The cell walls of Gram-positive and Gram-negative bacteria

The ultrastructure of eukaryotic cells

REVISED

Figure 2.3 shows the features of eukaryotic cells. The left-hand side shows features common to animal cells and the right-hand side shows features common to plant cells.

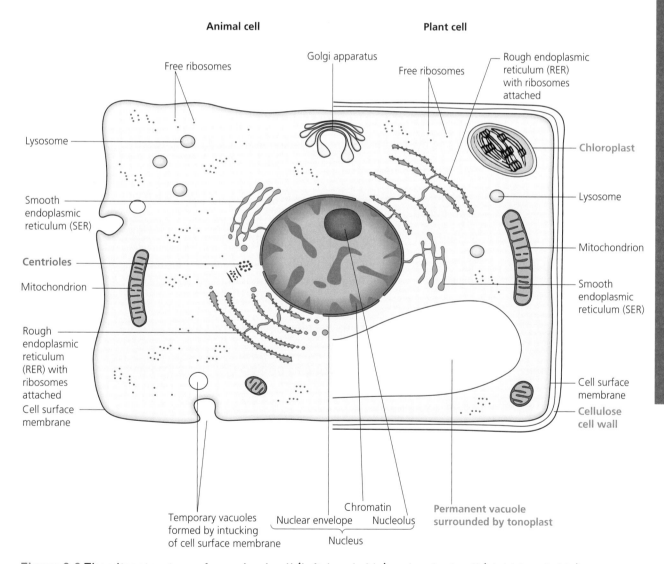

Animal cell

Plant cell

Free ribosomes

Golgi apparatus

Free ribosomes

Rough endoplasmic reticulum (RER) with ribosomes attached

Lysosome

Chloroplast

Smooth endoplasmic reticulum (SER)

Lysosome

Mitochondrion

Centrioles

Mitochondrion

Smooth endoplasmic reticulum (SER)

Rough endoplasmic reticulum (RER) with ribosomes attached

Cell surface membrane

Cell surface membrane

Cellulose cell wall

Temporary vacuoles formed by intucking of cell surface membrane

Nuclear envelope

Chromatin

Nucleolus

Permanent vacuole surrounded by tonoplast

Nucleus

Figure 2.3 The ultrastructure of an animal cell (left-hand side) and a plant cell (right-hand side)

Notice that many of the organelles within the cytoplasm are common to both and many are surrounded by membranes (i.e. are **membrane-bound organelles**). These include:

- tubules forming the endoplasmic reticulum
- flattened sacs and vesicles forming the Golgi apparatus
- lysosomes
- mitochondria and chloroplasts, both of which have an outer smooth membrane and an inner folded membrane

Table 2.2 summarises the functions of the cell components shown in Figure 2.3.

> **Exam tip**
>
> Students often save time by using abbreviations, for example ER for endoplasmic reticulum. Examiners expect you to use the terminology given in the specification, so only use abbreviations if they appear there.

Table 2.2 The main components of animal and plant cells

Cell component	Principal function(s)
Cell surface membrane	Controls the movement of ions and molecules into and out of the cytoplasm Contains proteins that act as receptors or as cell identification markers
Cell wall	Provides support for plant cell and resists pressure resulting from entry of water by osmosis
Centrioles	In animal cells, produce the spindle that separates copies of chromosomes during mitosis
Chloroplasts	Site of photosynthesis
Golgi apparatus	Modifies and packages polymers The vesicles that bud from the end of the flattened sacs of the main Golgi apparatus contain the packaged substances that will be used elsewhere in the cell or secreted from the cell
Lysosomes	Release hydrolytic enzymes (lysozymes) that digest unwanted material
Nucleolus	Site of ribosome synthesis
Nucleus	Contains the cell's chromosomes — DNA molecules each wound around protein called **histone**
Mitochondria	Produce ATP during aerobic respiration
Permanent vacuole	Maintains the turgor of plant cells, keeping the cytoplasm pushed against the cell wall and allowing cell elongation in the growing roots and stems Stabilizes pH of cytosol In many plants, contains the chemicals that produce toxins if a cell is damaged by a herbivore
Rough endoplasmic reticulum	Modifies and transports proteins produced by the ribosomes
Smooth endoplasmic reticulum	Synthesises lipids
Tonoplast	Controls the movement of water and ions between the cytoplasm and the contents of the permanent vacuole
80S ribosomes	Synthesise polypeptides

Typical mistake

Many students fail to gain credit by writing that mitochondria produce energy. Write either that mitochondria release energy or, better still, that mitochondria produce ATP.

Now test yourself

TESTED

1 How are the functions of the nucleus, rough endoplasmic reticulum and Golgi apparatus interrelated in a protein-secreting cell?
2 The permanent vacuoles of some plants contain chemicals that, when the tonoplast is broken, react with components in the cytoplasm to produce toxins. Suggest one advantage of this property to the plant.
3 Give two features common to both prokaryotic and eukaryotic cells.

Answers on p. 198

Studying cells

REVISED

Most cells are too small to see with the naked eye. To study them, you need to magnify them using a microscope. Even then, because cytoplasm is usually colourless, you see very little of the cell structure. A common way to overcome this is to **stain** cells before examining them with a microscope. The stains are selected to react with only one type of chemical within a cell, so that they are taken up selectively by different cell components. For example, as DNA is acidic, you can stain chromosomes with a basic dye.

Magnification and resolution

Magnification is a measure of how much larger an image of an object is than the actual object itself. The following formula shows how to calculate the magnification of an object:

$$\text{magnification} = \frac{\text{size of image}}{\text{size of object}}$$

Depending on which values you know, you can transpose this formula to calculate the magnification or the actual size of an object you are viewing with a microscope.

Resolution is a measure of the ability to distinguish between very close objects.

As a result of its wavelength, light cannot pass between objects that are less than about 0.2 µm apart. No matter how much you enlarge an image using a light microscope, the poor resolution of light will not allow you to see small objects clearly. Using an electron microscope overcomes this problem, as electrons have a much smaller wavelength than light. An electron microscope allows a resolution of 5 nm.

Table 2.3 summarises some features of light microscopes and electron microscopes.

Revision activity

Transpose the formula for calculating magnification to show how you could calculate:

(a) the size of an image

(b) the size of an object

Exam tip

You must be able to explain clearly the difference between magnification and resolution.

Typical mistake

Students often make mistakes when converting one unit to another. In terms of length, $1\,mm = 10^{-3}\,m$, $1\,\mu m = 10^{-3}\,mm$ and $1\,nm = 10^{-3}\,\mu m$. Bearing this sequence in mind, it is often best to take measurements of length in mm rather than in cm.

Now test yourself

TESTED

4 The scale bar on a photograph of a bacteria cell represents $2\,\mu m$. Using a ruler, a student measures this bar as 25 mm. What is the magnification of the photograph?

Answer on p. 198

Table 2.3 A comparison of light microscopes and electron microscopes

Feature	Light microscope	Electron microscope (EM)
Radiation source	Light from the environment or from a lamp	Electron beams fired from an electron 'gun'
How magnification and focusing are achieved	Glass lenses — condenser, eyepiece and objective	Electromagnets — condenser and objective
How the image is viewed	Looking down the eyepiece	Observing the image on a fluorescent screen
Types of stain	Chemical dyes that colour specific chemical within cells	Heavy metal ions that absorb electrons
Limitations	Resolution is much weaker than EM	Contains a vacuum (to prevent air deflecting the electrons) so it cannot be used to observe living cells

Viruses

The classification of viruses

REVISED

The features common to all viruses are:
- an outer protein coat, called a **capsid**
- a core of nucleic acid

The shape of the capsid and nature of the nucleic acid are used to classify viruses. Table 2.4 shows how this is done using four different viruses.

Table 2.4 Four types of virus

Name of virus	Host	Structure of capsid	Nature of nucleic acid
Lambda (λ) bacteriophage	*Escherichia coli*	Many-sided head, with tail and tail fibres	Double-stranded DNA
Tobacco mosaic virus	Plants	Spiral polypeptide	Single-stranded RNA
Ebola virus	Humans	Twisted tube	Single-stranded RNA
Human immunodeficiency virus (HIV)	Humans	Spherical	Single-stranded RNA

The lytic cycle and latency

Figure 2.4 shows the way in which one virus, the lambda (λ) bacteriophage, invades a cell of its specific host, the bacterium *Escherichia coli*. Because infection by the λ bacteriophage results in the bursting of the host cell, the cycle shown in Figure 2.4 is called the **lytic cycle**.

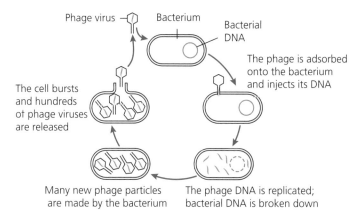

Figure 2.4 The lytic cycle of the λ bacteriophage

Latency refers to a period of time when a cell that is infected by a virus shows no symptoms. During this time, the nucleic acid of the virus has become incorporated into the DNA of the host cell, where it remains 'silent'. Later, stress results in the viral nucleic acid being translated by the host cell, which then begins to manufacture new viruses.

Viruses are not living cells

Viruses are not considered to be living cells because they lack a **metabolism**. A virus acts as an infective agent, using the metabolism of another cell (the host cell) to manufacture new virus particles. The host cell:

- replicates the viral nucleic acid
- translates the viral nucleic acid to produce new viral proteins
- assembles the viral nucleic acid and proteins to make new viruses

> **Metabolism** is the chemical reactions that occur in living cells and living organisms, including DNA replication, polypeptide synthesis and ATP hydrolysis.

Now test yourself

5 Explain why antibiotics have no effect on viruses.
6 Explain why the control of viral diseases, such as the 2014 Ebola outbreak in West Africa, focuses on prevention of their spread.
7 Cold sores are caused by viral infections, usually of the lips. Suggest why people who suffer cold sores often find they recur in the same place.

Answers on p. 198

Eukaryotic cell cycle and division

In a complex multicellular organism, cell division achieves three biological outcomes:
● growth from a single-celled zygote to a multicellular adult
● repair or replacement of damaged cells
● asexual reproduction

Following their formation in a complex multicellular organism, eukaryotic cells usually become specialised through a process of **differentiation**. In doing so, most lose the ability to divide and produce new cells.

The cell cycle

REVISED

Cells that retain the ability to divide undergo a controlled sequence of events called the **cell cycle**. This cycle, shown in Figure 2.5, involves:
● interphase
● mitosis
● cytokinesis

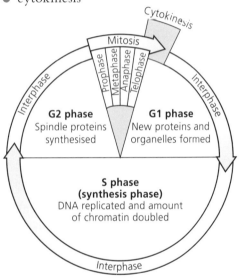

Figure 2.5 The eukaryotic cell cycle

> **Exam tip**
>
> Examiners will tolerate many spelling errors, but not when they relate to words with close spellings. You must spell 'mitosis' and 'meiosis' correctly.

Interphase

The time between divisions is a time of intense cellular activity:
● **G1 phase** — the cell enlarges after its last division and produces more organelles, nucleotides and histones.
● **S phase** — the cell replicates its DNA and combines it with newly synthesised histones so that it now holds two copies of each chromosome (called **sister chromatids**).
● **G2 phase** — growth of the cytoplasm continues and the cell manufactures more tubulin, the protein from which its spindle will be made.

Mitosis

Mitosis involves the division of the nucleus. The two copies of each chromosome (**chromatids**) are separated and grouped into two new nuclei.

Although mitosis is a continuous process, it is often described in the four discrete phases shown in Figure 2.6 and summarised in Table 2.5.

> A **chromatid** is one of the two copies of each chromosome formed after DNA replication and held together by a region called a centromere.

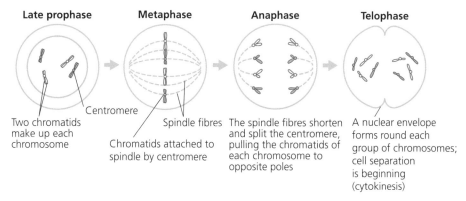

Late prophase | Metaphase | Anaphase | Telophase

Two chromatids make up each chromosome

Centromere

Chromatids attached to spindle by centromere

Spindle fibres

The spindle fibres shorten and split the centromere, pulling the chromatids of each chromosome to opposite poles

A nuclear envelope forms round each group of chromosomes; cell separation is beginning (cytokinesis)

Figure 2.6 Mitosis

Table 2.5 The events occurring during mitosis

Phase of mitosis	Summary of events
Prophase	The chromosomes shorten and thicken, each becoming visible as two sister chromatids held together at a point called the centromere
	The nuclear envelope disperses within the cytoplasm
	A spindle of tubulin fibres forms across the cell. In animals, the centrioles are involved in making this spindle
Metaphase	Contraction of spindle fibres pulls the pairs of sister chromatids by their centromere to the centre of the spindle
Anaphase	The centromeres divide, allowing separation of the two sister chromatids they hold together
	Further contraction of the spindle fibres pulls the sister chromatids apart, one of each pair to opposite poles of the spindle
Telophase	A new nuclear envelope forms around each group of chromosomes
	The chromosomes begin to elongate

Cytokinesis

The cytoplasm divides into two, almost equal, cells. In animal cells, this occurs by a pinching of the surface membrane from the outside to the centre, as shown in Figure 2.6. In plant cells, this occurs by production of a new cell wall from the centre to the outside of the cell.

Meiosis

REVISED

Meiosis is a second type of nuclear division that occurs in eukaryotes. It differs from mitosis in two important ways:

- It produces cells that contain half the number of chromosomes as the parent cell (**haploid cells**).
- It produces cells that are not genetically identical.

Unlike mitosis, meiosis involves two separate divisions. In the first division, **meiosis I**, **homologous chromosomes** are separated. In the second division, **meiosis II**, sister chromatids are separated. Figure 2.7 summarises the stages of each of these two divisions.

Exam tip

Describing the anaphase stage of mitosis, you could correctly write that 'the sister chromatids are separated' or 'the two copies of each chromosome are separated'.

Revision activity

Use four pencils and pieces of modelling clay to make a model of two chromosomes as they would appear in (a) prophase, (b) metaphase, (c) anaphase and (d) telophase of mitosis.

Homologous chromosomes are the two copies of each chromosome, one from each parent, in a zygote, that are copied to every cell by mitosis.

(a) The main stages of meiosis I

Early prophase I
The DNA has already replicated and each chromosome consists of two chromatids. The cell contains two sets of chromosomes.

Late prophase I
The spindle starts to form. Homologous chromosomes pair up forming bivalents and exchange DNA between non-sister chromatids.

Metaphase I
The spindle is complete and the nuclear envelope has disintegrated.The bivalents are arranged around the middle of the spindle.

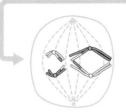  

Anaphase I
Homologous chromosomes start to separate from each other.

Telophase I
There is one complete set of chromosomes at each end of the cell.

Cytokinesis
The cytoplasm divides and two identical daughter cells are formed.The two cells resulting from meiosis I are therefore haploid.

> **Exam tip**
>
> Don't be so overawed by the process of meiosis that you forget its essential features — it produces cells that are genetically different and have half the chromosome number as the parent cell.

(b) The main stages of meiosis II

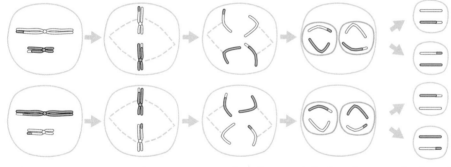

Prophase II
Each cell contains one chromosome from each homologous pair. The spindles begin to form.

Metaphase II
The chromosomes are aligned on the centre of the spindles.

Anaphase II
The chromatids are separated. Note: random segregation occurs here also.

Telophase II
Two haploid nuclei in each cell.

Cytokenesis
Four genetically different haploid cells.

Figure 2.7 The stages of the two divisions of meiosis

Genetic variation

At the end of meiosis, each homologous chromosome contains the same genes as its matching chromosome. These chromosomes do not, however, contain the same **alleles** of those genes, i.e. they show genetic variation. Two events that occur in meiosis contribute to the genetic variation in these offspring cells: independent assortment and crossing over.

> **Alleles** are two or more versions of the same gene, carrying slightly different DNA base sequences.

Independent assortment

During anaphase I, the direction in which each chromosome from a pair is pulled is completely random. Each new group, therefore, contains a different mixture of maternal and paternal chromosomes (represented by the two colours in Figure 2.7). Similarly, at anaphase II, the direction in which each sister chromatid from a pair is pulled is completely random.

Crossing over

In prophase I, non-sister chromatids in a homologous pair can become entangled at one or more points called **chiasmata**. This can result in breakage of the DNA strands. If the broken fragments rejoin the strand from which they came, there is no effect of the breakage. If the broken fragments join to the broken chromatid of the wrong chromosome, that DNA molecule has a new combination of the alleles of genes from where the breakages occurred. You can see this in Figure 2.7, where some individual DNA molecules are represented with both red and white components. Each new group, therefore, contains a mixture of chromatids in which one or more cross-overs occurred and those in which no cross-overs occurred.

> **Chiasmata** (singular: chiasma) are points at which non-sister chromatids become entangled and break during prophase I.

Now test yourself

TESTED ☐

8 Humans have 23 pairs of homologous chromosomes.
  (a) Assuming only independent assortment of homologous chromosomes occurs during sperm production, how many genetically different sperm cells could a man produce?
  (b) Assuming the same for the woman producing ova, how many genetically different children could a couple theoretically produce?

Answers on p. 198

Chromosome mutations

REVISED ☐

A chromosome mutation results in either:
- chromosome translocation — a change in the combination of genes on an individual chromosome
- chromosome non-disjunction — a change in the number of chromosomes

Chromosome translocation

Chromosome translocation causes a change in the combination of genes on a chromosome. Crossing over (see above) results from chromosome breakage. Sometimes, a broken fragment from a chromosome attaches to another chromosome that is neither its original chromosome nor its homologous chromosome. This is called **chromosome translocation**.

About 5% of cases of Down's syndrome in humans result from fertilisation involving an ovum in which a fragment from chromosome 21 has broken and joined chromosome 14.

Chromosome non-disjunction

Chromosome non-disjunction causes a change in the number of chromosomes. The chromosomes forming one homologous pair are usually separated during meiosis. If they fail to separate, one daughter cell will gain both copies of that pair and the other daughter cell will gain no copy of that pair. This is called **chromosome non-disjunction**.

Down's syndrome

Humans have 23 pairs of chromosomes, denoted 1 through 23. Down's syndrome is an inherited condition in humans resulting from non-disjunction of chromosome pair 21 during egg production. If an egg cell containing both copies of this chromosome is fertilised by a sperm cell with the usual one copy of the chromosome, a zygote is formed with three copies of chromosome 21. This condition is called **trisomy**. It can occur in other mammals too.

> **Trisomy** is the possession by a diploid cell of three copies of one of its chromosomes.

Turner's syndrome

Human females have two X chromosomes (represented XX) and males have one X chromosome and one shorter Y chromosome (represented XY). Turner's syndrome is an inherited condition in humans resulting from non-disjunction of the two X chromosomes during egg production. If an egg cell containing neither X chromosome is fertilised by a sperm cell with one copy of the X chromosome, a zygote is formed with only one sex chromosome (represented XO). This condition is called **monosomy**.

Now test yourself

TESTED

9 Most examples of monosomy are lethal. Suggest why Turner's syndrome is not.

Answer on p. 198

Sexual reproduction in mammals

The events that occur during sexual reproduction are basically the same in all mammals. You need to understand three:

- Gamete production (**gametogenesis**) — the formation of egg cells (**oogenesis**) and sperm cells (**spermatogenesis**).
- The events that occur when a sperm fertilises an egg cell.
- The development of the zygote to the blastocyst stage.

Oogenesis and spermatogenesis

REVISED

Figure 2.8 shows that the processes of oogenesis and spermatogenesis are basically similar.

Mitotic division of a germinal epithelial cell in the ovum or testis, followed by repeated mitotic divisions, produces vast numbers of **primary oocytes** and **primary spermatocytes**. The first meiotic division of each of these primary cells produces two haploid cells:

- A **secondary oocyte** and **polar body** (which is lost) in females.
- Two secondary spermatocytes in males.

> **Exam tip**
>
> Make sure you know where mitosis, meiosis I and meiosis II occur during gametogenesis.

The second meiotic division (of these secondary cells):

- in females occurs only after fertilisation and produces a single **ovum** and a second polar body (which is lost)
- in males produces four **spermatids** that mature into **sperm cells**

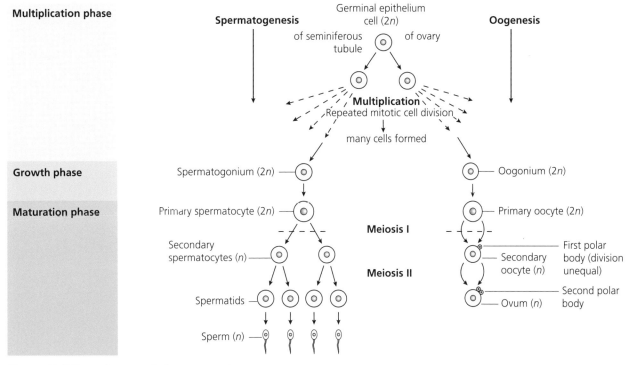

Figure 2.8 Gametogenesis in mammals

Fertilisation

REVISED ☐

Fertilisation involves:

- the fusion of one sperm with a secondary oocyte
- completion by the secondary oocyte of the second meiotic division to produce an ovum
- fusion of the haploid nuclei of the ovum and sperm to form a diploid cell, the **zygote**

The secondary oocyte

The term 'ovulation' is rather misleading. It is not an ovum that is released from an ovary at ovulation, but a secondary oocyte. Figure 2.9 shows the generalised appearance of a secondary oocyte at this stage. It is a large cell, surrounded by layers of cells from the structure in the ovary within which it was formed — the follicle.

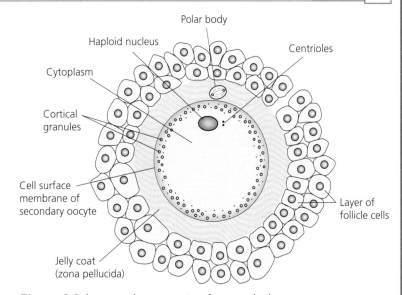

Figure 2.9 A secondary oocyte after ovulation

The collection of cells shown in Figure 2.9 is captured by the oviduct — a tube leading from the ovary to the uterus. It is moved along by the beating of cilia on the surface of cells lining the oviduct.

Exam tip

Take care to use the terms 'primary oocyte', 'secondary oocyte' and 'ovum' correctly.

TESTED

10 How is a secondary oocyte adapted for its function?

Answer on p. 198

The sperm

Figure 2.10 shows a mature sperm. Of the millions ejaculated into the female's vagina, only a few ever reach the secondary oocyte and, of these, only one can fertilise it. A sperm is well adapted for fertilisation by being very small and possessing:
- a tail, enabling it to swim
- a middle piece packed with mitochondria that produce the ATP used in swimming
- an **acrosome** containing hydrolytic enzymes

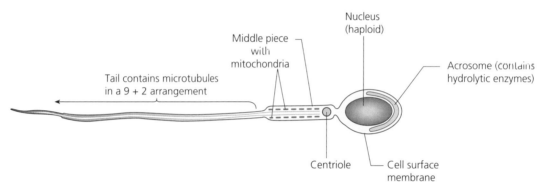

Figure 2.10 **A mature sperm**

Fertilisation and the cortical reaction

Figure 2.11 summarises the events that occur at fertilisation. Notice the importance of:
- the hydrolytic enzymes of the sperm head that digest a way to the cell surface membrane of the secondary oocyte
- the **cortical reaction**, in which granules from the cortex (outer region) of the secondary oocyte form a barrier to prevent the entry of another sperm
- stimulation of the secondary oocyte to complete meiosis, forming an ovum

Development of the zygote into a blastocyst

REVISED

After fertilisation, the diploid zygote begins to divide as it is moved along the oviduct to the uterus. A series of mitotic divisions produces a 2-cell stage, a 4-cell stage, an 8-cell stage, and so on.

In humans, 7 days after fertilisation the initial solid ball of cells has developed into a hollow ball of 128 cells. This is the **blastocyst**, shown in Figure 2.12. Even at this stage, the blastocyst has begun to differentiate.

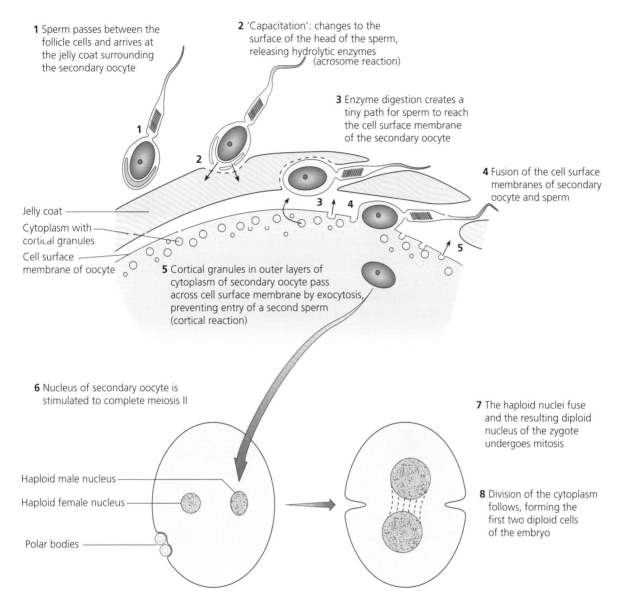

1 Sperm passes between the follicle cells and arrives at the jelly coat surrounding the secondary oocyte

2 'Capacitation': changes to the surface of the head of the sperm, releasing hydrolytic enzymes (acrosome reaction)

3 Enzyme digestion creates a tiny path for sperm to reach the cell surface membrane of the secondary oocyte

4 Fusion of the cell surface membranes of secondary oocyte and sperm

Jelly coat

Cytoplasm with cortical granules

Cell surface membrane of oocyte

5 Cortical granules in outer layers of cytoplasm of secondary oocyte pass across cell surface membrane by exocytosis, preventing entry of a second sperm (cortical reaction)

6 Nucleus of secondary oocyte is stimulated to complete meiosis II

7 The haploid nuclei fuse and the resulting diploid nucleus of the zygote undergoes mitosis

8 Division of the cytoplasm follows, forming the first two diploid cells of the embryo

Haploid male nucleus

Haploid female nucleus

Polar bodies

Figure 2.11 Fertilisation of a secondary oocyte by a sperm

Cells of the inner cell mass, the **blastomeres**, will form the fetus, which will become the newborn mammal. Cells of the outer layer, the **trophoblasts**, will form the membrane that surrounds the fetus while it is in the uterus. The fluid in the blastocoel will protect the fetus from mechanical shock throughout its development and will be lost at birth as the 'broken waters'.

About 14 days after fertilisation, a successful human blastocyst will have become embedded in the thickened uterus lining. Its trophoblast will have started to develop finger-like extensions that grow into the lining of the uterus.

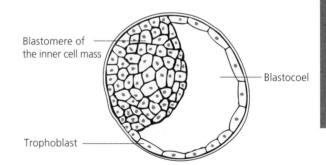

Blastomere of the inner cell mass

Blastocoel

Trophoblast

Figure 2.12 A 7-day-old blastocyst

Now test yourself

TESTED

11 How many cell divisions will have occurred between formation of the zygote and the 7-day-old blastocyst?
12 Suggest two advantages to the blastocyst of its trophoblast forming finger-like projections into the lining of the uterus.

Answers on p. 198

Sexual reproduction in plants

The reproductive organs of a flowering plant are enclosed within its flowers. Like many flowering plants, the buttercup shown in Figure 2.13 has both female and male organs within each of its flowers. The female organ is inside a **carpel**, consisting of a **stigma**, **style** and an **ovary** that contains an **ovule**. The male organ is inside a **stamen**, consisting of a **filament** and an **anther** that contains **pollen grains**.

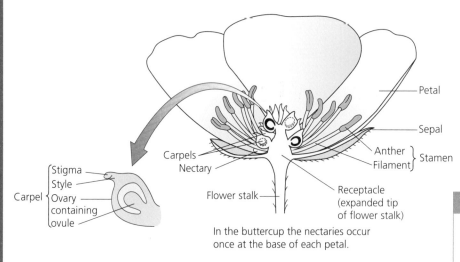

In the buttercup the nectaries occur once at the base of each petal.

Figure 2.13 A vertical section through a buttercup flower

Unlike those in mammals, the gametes of a flowering plant are not cells; they are haploid nuclei within a spore.

> **Revision activity**
>
> Take a large, simple flower, such as a tulip or iris. Using a scalpel and forceps, carefully dissect the plant to find the structures shown in Figure 2.13.

The formation of female gametes

REVISED

Within the large central mass of the ovule is a cell called the **megaspore mother cell**. This divides by meiosis to produce four haploid **megaspores**. Three of the megaspores degenerate. The fourth enlarges to become the single-celled **embryo sac**. The haploid nucleus within the embryo sac divides by mitosis to form eight haploid nuclei. All these nuclei remain within the single-celled embryo sac; its cytoplasm does not divide.

- One of these nuclei in the embryo sac is the **female gamete**.
- Two of these nuclei in the embryo sac are the **polar nuclei**.

You can see these nuclei in Figure 2.15.

The formation of male gametes

Each anther has a number of lobes. Within the tissue of these lobes are many **microspore mother cells**. Each microspore mother cell divides by meiosis to produce four haploid **microspores**. The haploid nucleus within each microspore then divides once by mitosis to produce two haploid nuclei.
- One is the generative nucleus.
- The other is the pollen tube nucleus.

These two nuclei remain within what is now called a single-celled **pollen grain**; its cytoplasm does not divide.

> **Exam tip**
>
> Don't refer to pollen grains as gametes; the gametes are haploid nuclei within the pollen grains.

Now test yourself

13 Give one way in the gametes of a mammal differ from those of a flowering plant.

Answer on p. 198

Double fertilisation

Fertilisation of a flowering plant involves:
- mitotic division of the generative nucleus of a pollen grain to produce two haploid male gametes
- fusion of one male gamete with the female gamete inside the embryo sac
- fusion of one male gamete with both polar nuclei inside the embryo sac

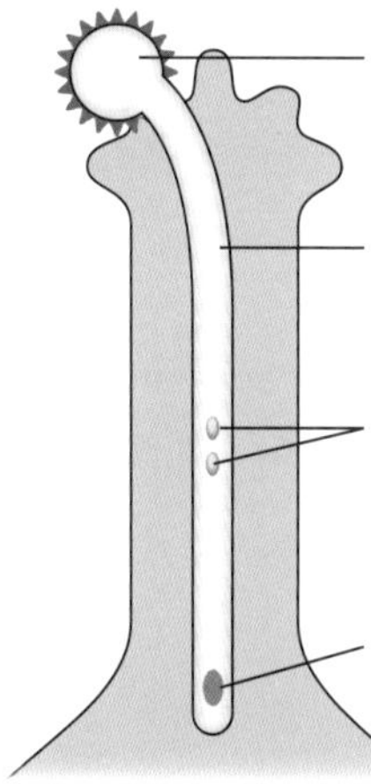

Pollination occurs when a pollen grain lands on the stigma of a flower of the same species. If the match between the stigma and the pollen grain is suitable, the pollen grain absorbs water from the stigma, swells and bursts. The burst pollen grain produces a **pollen tube** that, under the control of the pollen tube nucleus, grows through the tissue of the style to the embryo sac. The generative nucleus follows the pollen tube nucleus down the pollen tube. As it does so, this nucleus divides once by mitosis to form two nuclei — the **male gametes** (Figure 2.14).

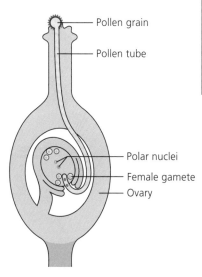

Figure 2.14 **Growth of a pollen tube and production of two male gametes**

— Pollen grain

— Pollen tube

— Male gametes

— Pollen tube nucleus

> **Exam tip**
>
> Remember that mitosis produces daughter cells with the same chromosome number as the parent cell. If the parent cell is haploid, so are the daughter cells produced by mitosis.

Once inside the embryo sac, the tip of the pollen tube breaks down and the pollen tube nucleus disintegrates (Figure 2.15). One male gamete fuses with the female gamete, forming a diploid **zygote**. Mitotic divisions of this nucleus will produce an embryo plant. The other male gamete fuses with both polar nuclei, forming a triploid **primary endosperm cell**. Mitotic divisions of this nucleus will produce endosperm — a fuel store for the developing embryo.

As the embryo and endosperm grow and develop, changes also occur in the female organs. The embryo sac develops in a **seed**. The ovary wall develops into a **fruit** that aids in the dispersal of the mature seed.

— Pollen grain

— Pollen tube

— Polar nuclei

— Female gamete

— Ovary

Figure 2.15 **Double fertilisation**

Now test yourself

14 Distinguish between pollination and fertilisation.
15 During the growth of a pollen tube, suggest:
   (a) the function of the pollen tube nucleus
   (b) the source of the raw materials for growth of the tube.
16 Tomatoes and cucumbers are often described as vegetables. In fact, they are fruit. Explain why.

Answers on pp. 198–199

Exam practice

1 Complete the table by putting a tick (✓) in each empty cell in which the feature is present in the type of cell and a cross (✗) where it is not present. [6]

Feature	Animal cell	Bacterial cell	Plant cell
Cell wall			
Cell surface membrane			
Chloroplast			
Mitochondria			
Nucleoid			
Tonoplast			

2 The diagram represents four stages in the division of a cell. They are not shown in the correct order.

A    B   C   D

(a) Put the letters **A** to **D** into the order in which they would occur during division of this cell. Start with **D**. [1]
(b) Name the type of nuclear division that is shown in the diagram. Justify your answer. [2]
(c) Describe the role of the structure labelled **X**. [2]

3 The drawing was made from an electron micrograph of a cell from the pancreas of a rat.

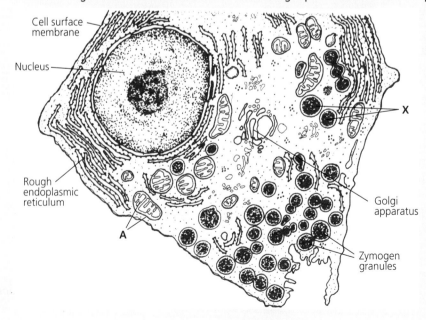

(a) What is the evidence that the drawing was from an electron micrograph? [1]

(b) Name the two structures labelled **A**. [1]

(c) Explain why the two structures labelled **A** are not the same shape. [1]

(d) The zymogen granules are full of inactive enzymes. Explain how this cell is adapted to secrete large amounts of these enzymes. [3]

(e) The actual diameter of the zymogen granule labelled **X** is 1.2 μm. Calculate the magnification of this drawing. Show your working. [2]

4 A scientist investigated the effect of sucrose concentration on the growth of pollen tubes in two species of tree: apricot and sweet cherry.

He inoculated pollen grains of each species onto agar plates containing different concentrations of sucrose. He then incubated the agar plates in the dark at 22°C. After 24 hours, he found the mean length of 100 pollen tubes from each agar plate. His results are shown the table.

Concentration of sucrose solution/%	Mean length of pollen tube/μm	
	Apricot	Sweet cherry
5	267	233
10	295	239
15	282	293
20	301	249

(a) Given a 20% solution of sucrose, describe how you would produce 20 cm³ of a 5% sucrose solution. [1]

(b) What is the evidence from the table that the scientist rounded the value of his calculated means? [2]

(c) Plot a suitable graph of the data in the table. [4]

(d) Describe what your graph shows about the effect of sucrose concentration on pollen tube growth. [3]

5 (a) The table shows the mass of DNA in some mammalian cells.

Mammalian cell	Mass of DNA/arbitrary units
Muscle cell	36
Red blood cell	0
Germinal epithelial cell in the G2 phase of the cell cycle	
Sperm cell	
Primary oocyte	

(i) Explain the mass of DNA in a red blood cell. [1]

(ii) Complete the table to show the mass of DNA in the germinal epithelial cell in the G2 phase of the cell cycle, sperm cell and primary oocyte. [3]

(b) Explain why meiosis results in genetic diversity but mitosis does not. [6]

6 A bacteriophage is a virus that infects bacteria. The diagram shows a bacteriophage that causes a lytic cycle in the bacterium *Escherichia coli*.

Protein ⎯

DNA ⎯

(a) Describe one way visible in the diagram in which the bacteriophage is:

(i) similar to a living organism [1]

(ii) different from a living organism. [1]

(b) Describe the lytic cycle caused by this bacteriophage. [4]

(c) A team of scientists used radioactive elements to label two different types of molecule in a culture of this bacteriophage. The radioactive isotopes they used were ³²P and ³⁵S.

They then infected cells of *E.coli* with the bacteriophage. After a few minutes, they separated the bacteriophage and the cells of *E. coli*. Finally, they investigated where the radioactivity was located. The table summarises their results.

Structure	Location of radioactivity	
	Radioactive phosphorus (^{32}P)	Radioactive sulfur (^{35}S)
Bacteriophage	✗	✓
Cells of *E. coli*	✓	✗

(i) Name the type of biological molecule that contains phosphorus and the type of biological molecule that contains sulfur. [2]

(ii) Explain the scientists' results shown in the table. [2]

> **Exam tip**
>
> Question 6 tests content from more than one topic in the specification. This is called **synopsis** and is common in the exams. In preparing for your exam, ensure that you make links between different, but related, topics.

Answers and quick quiz 2 online

ONLINE

Summary

Eukaryotic and prokaryotic cell structure and function

- The cell theory is a unifying concept in biology.
- In complex multicellular organisms, specialised cells are organised into tissues, tissues into organs, and tissues and organs into systems.
- A prokaryotic cell is much smaller than a eukaryotic cell and lacks a nucleus and membrane-bound organelles. All prokaryotic cells have a cell wall, a cell surface membrane and cytoplasm containing a single nucleoid, 70S ribosomes and food granules.
- A eukaryotic cell contains a nucleus and its cytoplasm is filled with membrane-bound organelles. All eukaryotic cells have a nucleus, 80S ribosomes, endoplasmic reticulum (rough and smooth), mitochondria, lysosomes and Golgi apparatus. Features of plant cells also include a cell wall, chloroplasts and a permanent vacuole surrounded by a tonoplast.
- Light microscopes and electron microscopes both allow objects to be magnified. The shorter wavelength of electrons enables electron microscopes to achieve much better resolution than light microscopes.

Viruses

- Viruses are non-living protein capsids surrounding a nucleic acid core. They invade specific host cells, which use the viral DNA to manufacture new virus particles.

- Viruses are classified on the basis of the structure of their capsid and nature of their nucleic acid.

Eukaryotic cell cycle and division

- Some eukaryotic cells undergo a regulated cell cycle involving interphase, mitosis and cytokinesis.
- Mitosis is a nuclear division consisting of four phases — prophase, metaphase, anaphase and telophase — during which the two copies of each replicated chromosome are separated into two new nuclei. Thus, a single mitotic division produces two nuclei that are genetically identical.
- Meiosis is a nuclear division that results in daughter cells that have half the chromosome number as the parent cell and are genetically different from each other.
- The genetic variation between the daughter cells produced by meiosis results from:
  - independent assortment of homologous chromosomes during meiosis I
  - crossing over between homologous chromosomes during meiosis I
- Meiosis involves two nuclear divisions:
  - In meiosis I, the two chromosomes in each homologous pair are separated into two new nuclei.
  - In meiosis II, the two replicated copies of each chromosome (chromatids) are separated into new nuclei.

- Chromosome mutations include:
  - chromosome translocation, in which part of a chromosome breaks away and joins another chromosome
  - chromosome non-disjunction, in which two homologous chromosomes fail to separate during meiosis I. After fertilisation, non-disjunction results in zygotes that show trisomy or monosomy.

Sexual reproduction in mammals

- Mammalian gametes are produced by oogenesis and spermatogenesis. In each, a germinal epithelial cell divides by mitosis to form primary oocytes and primary spermatocytes. These cells undergo meiosis to form mature gametes.
- Fertilisation involves the successful fusion of a male gamete nucleus with a female gamete nucleus. In mammals, hydrolytic enzymes within the acrosome of its head enable the sperm to pass its nucleus into the cytoplasm of the secondary oocyte. The cortical reaction of the secondary oocyte prevents the entry of further sperm.
- Mitotic cell divisions of a mammalian zygote produce a hollow ball of cells — the blastocyst — that embeds in the thickened uterus wall of the female.

Sexual reproduction in plants

- In flowering plants, the gametes are nuclei held within spores: the female embryo sac and the male pollen grain.
- The male gametes of flowering plants reach the female gametes through a pollen tube that grows from the stigma to the embryo sac within a flower.
- Within the embryo sac, double fertilisation occurs. One male nucleus fuses with two female nuclei to form a triploid endosperm cell; the other male nucleus fuses with the female gamete nucleus to form a zygote.

3 Classification and biodiversity

Classification

Living organisms can be classified in two ways:
- An **artificial classification system** is based on how organisms affect humans, e.g. it is/is not good to eat or it will/will not attack.
- A **natural classification system** attempts to show the evolutionary relationships between organisms.

Taxonomy: the classification of living organisms REVISED

As with books in a library, organisms are classified by putting them into groups, called **taxa**. These groups are placed in a **hierarchy**. Organisms in the largest group in the hierarchy show fewer similarities than those in the smallest group in the hierarchy. There is no overlap between taxa at the same level within the hierarchy.

The smallest taxon in our classification system is the **species**. Table 3.1 shows the name of each major taxon with which you should be familiar. It also includes the names of the taxa that are relevant to humans and to the common oak tree, but you do not need to learn these names.

> **Taxa** (singular: taxon) are groups of organisms within the hierarchy of the biological classification system.
>
> A **species** is a group of organisms with similar characteristics that interbreed to produce fertile offspring.

Table 3.1 The major taxa used in taxonomy

Example using common oak tree	Name of taxon (largest first)	Example using humans
Eukarya	Domain	Eukarya
Plantae (plants)	Kingdom	Animalia (animals)
Angiospermophyta	Phylum	Chordata
Dictoyledonae	Class	Mammalia
Fagales	Order	Primata
Fagaceae	Family	Hominidae
Quercus	Genus	*Homo*
robur	Species	*sapiens*

> **Exam tip**
>
> It is often helpful to have a mnemonic to remember lists of terms. Using the initial letters of the taxa in Table 3.1, you could use the mnemonic **D**o **K**eep **P**ot **C**lean **O**r **F**amily **G**ets **S**ick.

Binomial nomenclature

Different languages throughout the world have different common names for every organism. To avoid confusion, scientists have adopted the system of binomial nomenclature, devised by Carl Linnaeus in the eighteenth century. It is based on the information shown in Table 3.1.

We need to use just two names — the genus and species — to give every organism a unique name. There are rules to this nomenclature system:
- The name of the genus is a noun; the name of the species is an adjective.
- The names are usually in Latin — a language no one uses today, so they are less likely to be corrupted.

- The name of the genus starts with an upper case letter whereas the name of the species starts with a lower case letter.
- The binomial is written in italics (or, when handwriting, underlined).
- So, a human is *Homo sapiens* and the oak tree a human might climb is *Quercus robur*.

Now test yourself

TESTED

1 The international system for naming organisms uses Latin. Give one advantage of this.
2 The European robin is *Erithacus rubecula* whereas the North American robin is *Turdus migratorius*. What can you conclude from this information?

Answers on p. 199

Exam tip

The examiner can test your ability to recall the taxon names in the middle column of Table 3.1. Your ability to recall the names of any specific organism will not be tested.

The limitations of the definition of a species

REVISED

In the natural classification of organisms, the smallest group we recognise is the species. Despite this being the fundamental group in our classification system, difficulties often arise when assigning organisms to a species.

Variation within a species is normal. How much variation can be tolerated before biologists classify organisms into different species? Differences between species are often not obvious. For example, there are about 1000 species of the fruit fly (*Drosophila*) on the islands of Hawaii, most of which you would find hard to tell apart — a vast number. To date, about 2×10^6 species have been described and named. Many new species are still being discovered and scientists do not know how many species are still unknown.

Since Carl Linnaeus devised the basis of the modern classification system still in use today, different fields of biology have evolved and, with them, different definitions of the term 'species' (Table 3.2). It is not unusual

Table 3.2 Different approaches to defining the term 'species', in approximate chronological order

Field of study	How the term 'species' could be defined in each field of study
Prior to scientific investigation to the present day	The direct descendants of the first pair that God created in the Garden of Eden
Anatomy	Differences and similarities of body shape and structure
Physiology	Differences and similarities in the way that tissues, organs and systems operate and are controlled
Ecology	Differences in the biological and physicochemical factors that determine where a group of organisms is able to survive, i.e. its ecological niche
Genetics	The ability of organisms to interbreed and produce fertile offspring
Biochemistry	Differences and similarities in the chemicals and chemical reactions that occur in cells
Immunology	The extent to which differences in the proteins of one organism cause reactions with antibodies from another organism
DNA technology and bioinformatics	Differences and similarities in nucleic acid base sequences and amino acid sequences of the polypeptides they encode

for two different techniques to suggested slightly different evolutionary relationships of the same groups of species.

Now test yourself

TESTED

3 Suggest two practical problems that might be encountered in applying the geneticist's definition of species given in Table 3.2.

Answer on p. 199

The contribution of DNA technology to classification

REVISED

The biological classification system seeks to establish evolutionary relationships between groups of organisms. The relatively new field of DNA technology has made a significant contribution to establishing these evolutionary relationships. The following summary covers the information you need to understand.

Now test yourself

TESTED

4 Is the biological classification system an artificial system or a natural system? Explain your answer.

Answer on p. 199

DNA hybridisation

This technique allows scientists to investigate the similarity between the DNA molecules of two organisms. Figure 3.1 summarises the stages in the technique.

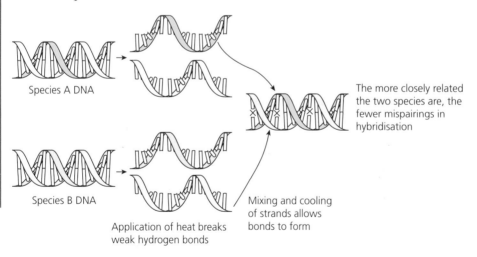

Species A DNA

Species B DNA

Application of heat breaks weak hydrogen bonds

Mixing and cooling of strands allows bonds to form

The more closely related the two species are, the fewer mispairings in hybridisation

Figure 3.1 DNA hybridisation is used to find similarities in DNA base sequences

DNA samples from two different organisms are heated in different tubes. This separates the DNA strands by breaking the hydrogen bonds between complementary base pairs. The two samples are mixed and allowed to cool. This allows the separate strands to hybridise, i.e. hydrogen bonds form between complementary base pairs on the different strands. The

hybridised strands are reheated until they separate again. The fewer the number of hydrogen bonds between them:
- the lower the temperature at which they separate
- the less similar their DNA base sequences
- the less closely related the organisms

DNA sequencing

As its name suggests, DNA sequencing determines the sequence of bases in a DNA molecule. The details of sequencing are complex, but the following description is all you need for this topic.

DNA is isolated from the test organism. The isolated DNA is hydrolysed into single-stranded fragments of different lengths. The single-stranded fragments are mixed with DNA nucleotides and DNA polymerase so that new strands of DNA are produced. Some of the nucleotides are slightly different from normal and carry a fluorescent label. When one of the labelled nucleotides is included in a developing DNA strand, it stops any more nucleotides being added. The result is a mixture of strands of different length, each with a fluorescent label that shows the last base that was added to the sequence (Figure 3.2).

Figure 3.2 **A mixture of single-stranded DNA, each of which has a fluorescent marker showing the last base that was added to it**

Now test yourself

TESTED

5 What is the function of the DNA polymerase used in the DNA sequencing technique?
6 The labelled nucleotides used in DNA sequencing have a hydrogen atom in the place of the hydroxyl group on their carbon-3. Suggest why inclusion of one of these labelled nucleotides stops further nucleotides being added to a developing strand.

Answers on p. 199

The strands in this mixture are separated using a technique called **electrophoresis**. In automated DNA sequencing machines, the process is carried out using gel within a capillary tube and laser scanners. DNA fragments carry a negative charge. When placed in an electric field, they migrate through the gel in a capillary tube to the positive electrode. The smaller fragments move through the gel faster and, therefore, further than the larger fragments. Lasers detect the fluorescent markers as they pass through the gel in the capillary, allowing identification of the DNA base sequence (Figure 3.3). Overlaps in the sequences of the fragments allow scientists to deduce the base sequence of the entire DNA molecule.

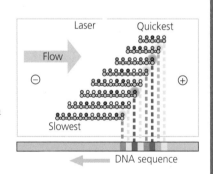

Figure 3.3 **Lasers detect the order of labelled DNA fragments separated by gel electrophoresis, allowing the DNA base sequence to be determined**

Now test yourself

TESTED

7 Suggest why DNA sequencing is carried out using DNA fragments rather than whole DNA molecules.
8 Separation of DNA by electrophoresis depends on DNA having a negative charge. Which component of DNA carries this negative charge?

Answers on p. 199

Bioinformatics

The automation of DNA sequencing has led to vast arrays of information about DNA base sequences.

Bioinformatics is the term used to describe the storage and indexing of this electronic information for future analysis and use. This relatively new branch of science combines the disciplines of applied mathematics, statistics and computer science.

DNA sequencing and bioinformatics enable comparisons to be made of the entire **genomes** of different organisms. Scientists can then use these comparisons to deduce how many gene mutations would be required to cause the observed differences in these genomes. Knowing the expected frequency of random gene mutations, scientists can then put a time line to these mutations. In this way, the divergence of two species can be estimated and a **phylogenetic tree**, like the one in Figure 3.4, created.

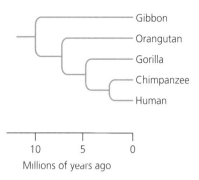

Figure 3.4 **A phylogenetic tree of the great apes (Hominidae) based on DNA evidence**

A **genome** is the complete set of DNA in each organism.

A **phylogenetic tree** is a diagram that attempts to show when different species diverged from a common ancestor.

How scientists come to an agreement about classification of species

REVISED

The way in which scientists agree about the classification of species is basically the same as the way scientists agree about any concept.

Underlying all scientific work is the **scientific method of enquiry** (Figure 3.5). As long as scientists around the world can produce experimental evidence that supports a particular theory, they become more confident that the theory is true. Of course, it only takes one experiment that fails to support a theory to make scientists question their beliefs.

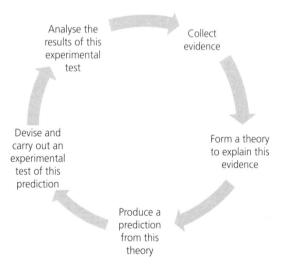

Figure 3.5 **The scientific method of enquiry**

Typical mistake

Students often write that the results of an experiment 'prove' a theory to be true. This is not the case; they can only support or fail to support a theory.

Exam tip

When asked to evaluate a scientist's results, always assume the scientist has worked professionally. Don't include statements that suggest otherwise, such as 'he might not have carried out enough repeats'.

Scientists seldom work in isolation. Rather, scientists in different laboratories around the world, or at different times, test predictions from the same theories. Like you, they write up reports of their investigations. These reports are published in **scientific journals**, enabling others to read them. Before a report is accepted for publication in a journal, it is sent to other scientists working in the same field for **peer review**. The reviewers check that, for example, the methodology used in the report was valid and the conclusions are justified by the results. Summaries of these reports are also often presented at **scientific conferences**. Here, other scientists working in the same field have the opportunity to

Peer review is the strict process through which a paper submitted for publication in a scientific journal is validated by independent experts in the same field before being accepted for publication.

question the researchers face to face. Through face-to-face discussions at conferences, articles in journals and the collaboration between groups that often follows, the scientific community **validates** the work of scientists and the evidence they have collected.

Now test yourself

TESTED

9 Explain why the scientific method can disprove a theory but not prove it.
10 Explain why peer review must be very strict.

Answers on p. 199

> **Exam tip**
>
> When describing measurements made during an investigation, use the terms **accurate** to mean 'close to the true value' and **precise** to mean 'there is little variation between repeated measurements of the same variable'.

Models of biological classification

REVISED

Currently, there are two major classification schemes in use, although they are not mutually exclusive:

● The **five-kingdom model** is based on structural differences between simple and more complex cell types and body forms.
● The **three-domain model** is more recent and is based on the results of DNA sequencing.

The five-kingdom model

The five kingdoms in this model are shown in Table 3.3. The table also summarises a few of the characteristic features of each kingdom.

Table 3.3 **Major features of the five kingdoms**

Name of kingdom	Characteristics
Prokaryotae	Possess prokaryotic cells Includes **autotrophs** and **heterotrophs**
Protoctista	Possess eukaryotic cells Body forms include unicells, colonies, filaments or complex body forms Includes autotrophs (e.g. algae) and heterotrophs (e.g. protozoa)
Fungi	Possess eukaryotic cells with cell walls made of chitin May be unicellular, but usually the body is a **mycelium** — a mass of threads called **hyphae** All are heterotrophs, feeding as **saprobionts** or **parasites** Motile cells never occur in the life cycle
Plantae	Possess eukaryotic cells with cellulose cell walls Multicellular with branching body growing from special patches of dividing cells (meristems) Autotrophs
Animalia	Possess eukaryotic cells with no cell wall Multicellular All are heterotrophs Develop from a blastocyst (see Figure 2.12)

> An **autotroph** is an organism that is able to synthesise complex organic compounds from simple inorganic compounds using an external energy source.
>
> A **heterotroph** is an organism that feeds on other organisms.
>
> A **saprobiont** is an organism that feeds by secreting digestive enzymes into its environment and absorbing the digested products.
>
> A **parasite** is an organism that lives in or on another living organism (its host), causing it damage.

11 Why would an autotroph use an external energy source when synthesising organic compounds from inorganic molecules?

Answer on p. 199

You can see that some of the kingdoms in Table 3.3 are easily recognised. A member of the kingdom Prokaryotae has a prokaryotic cell. However, this does not apply to the kingdom Protoctista. Members of this kingdom are mainly defined by exclusion: if an organism is not a member of the other four kingdoms, it is a protoctist.

> **Revision activity**
>
> Devise a table to summarise the features of the five kingdoms. Design your table so you can simply put ticks or crosses in each section of the table.

The three-domain model

An alternative classification model has been devised that takes account of research into the nucleotide sequence of the small unit of ribosomal RNA (known as 16S rRNA) and of bacterial genomes. This is the three-domain model, which is summarised in Table 3.4.

Table 3.4 **Major differences between the three domains (L-glycerol and D-glycerol are isomers of glycerol)**

Feature	Domain		
	Archaea	**Bacteria**	**Eukarya**
Cell structure	Prokaryotic	Prokaryotic	Eukaryotic
Nature of DNA	Circular	Circular	Linear (chromosomes)
DNA bound to protein (histone)	No	No	Yes
Presence of introns	Absent	Absent	Present between exons
Cell wall	Yes (but not peptidoglycan)	Yes (peptidoglycan)	In some (either cellulose or chitin)
Lipids in cell surface membrane	Branched fatty acids attached to L-glycerol by ether linkages. (The ether linkages enable them to live in extreme habitats, e.g. hot water springs.)	Unbranched fatty acids attached to D-glycerol by ester bonds	

12 L-glycerol and D-glycerol are isomers of glycerol. Explain what this means.
13 The five-kingdom model and the three-domain model are not mutually exclusive. Explain this statement.

Answers on p. 199

Natural selection

A fundamental principle in biology is that species can and do change. The process through which this occurs is called **natural selection**.

Adaptations

REVISED

Natural selection operates on the individuals of all species. The process of natural selection can be summarised as follows.

Intraspecific competition occurs in every population because the individuals within a population produce more offspring than can be supported by their environment. **Genetic variation** occurs in every sexually reproducing population as a result of mutations, the events of meiosis and random fertilisation. This genetic variation exists as **alleles** of the same genes. Some of this genetic variation gives individuals a **competitive advantage** over others in the population, which leads to **differential reproductive success**. In other words, those individuals carrying the genetic advantage have more offspring than those that do not carry the genetic advantage. The genetic variation that causes the competitive advantage is passed on to offspring, so more individuals in the next generation carry the alleles of genes that confer a competitive advantage. In other words, the frequency of the advantageous alleles increases in each generation. This is **evolution**.

> **Evolution** is a long-lasting change in the frequency of the alleles of a single gene within a population.

> **Exam tip**
>
> Avoid using the term 'survival of the fittest'. It is likely to be ignored by examiners because it shows no understanding on your part of either competitive advantage or of differential reproductive success.

Now test yourself

TESTED

14 What causes genetic variation in populations of asexually reproducing organisms?
15 Give two reasons why an individual with a competitive advantage might have more offspring than another without the competitive advantage.

Answers on p. 199

Directional and stabilising selection

In the explanation above, the frequencies of alleles of a single gene are changing generation after generation. This is called **directional selection**.

In many stable populations, individuals are well adapted. The most advantageous characteristic is the most common one. Any deviation from this optimum puts its carrier at a selective disadvantage. Natural selection that favours the existing adaptation at the expense of deviations from it is called **stabilising selection**.

Both types of selection are shown in Figure 3.6.

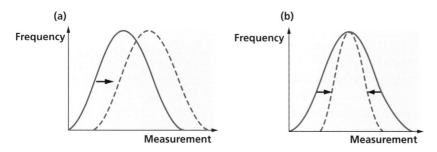

Figure 3.6 The effects of (a) directional and (b) stabilising selection

Ecological niches

Members of a species are restricted in where they can survive. Each species is limited by those aspects of its environment that it can tolerate. These include:

● physical and chemical factors, such as temperature range or pH of soil
● biological factors, such as the presence of competitors for the same resource

If we could take account of all the factors that affect the ability of a species to survive, we would describe its **ecological niche**. Figure 3.7 represents the tolerance ranges of a hypothetical species to three environmental factors. This species cannot survive successfully outside any one of these tolerance ranges. Consequently, it can only survive where all three overlap. This begins to describe its ecological niche. If we were able to construct a three-dimensional graph with axes representing all the environmental factors affecting this species, we would describe its true ecological niche.

> An **ecological niche** is a description of where a species can successfully exist that includes all its interactions with the living and non-living environment.

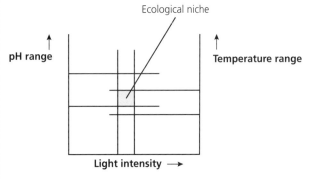

Figure 3.7 A model of the ecological niche of a hypothetical species is described as where all of its tolerance ranges overlap

Members of a species survive in their niche so long as they are better adapted to that niche than members of any other species. If another species is better adapted to that niche it can, through competition, exclude the existing species. These adaptations include:

● **physiological adaptations**, e.g. plants that grow in alpine meadows have enzymes with lower optimum temperatures than those that inhabit the tropics
● **behavioural adaptations**, e.g. remaining dormant during times when conditions deteriorate
● **anatomical adaptations**, e.g. humans that originated in a cold environment, such as the Inuit people of Greenland, have a body shape with a lower surface area to volume ratio than that of humans originating in hot plains, such as the Maasai people of Tanzania and Kenya

> **Typical mistake**
>
> Students often write statements such as 'cacti have developed short spiny leaves to adapt to dry conditions'. Such statements incorrectly imply that organisms have a purpose and can consciously change to achieve that purpose. It is better to write that 'the short, spiny leaves of cacti are an adaptation that enables them to withstand desiccation in dry conditions'.

Reproductive isolation

Allopatric speciation

Figure 3.8 shows what can happen if a single population is split into two by a geographical barrier. The two groups have now become reproductively isolated. In the different environments, natural selection can act differently on the individuals in the two populations. As a result, the allele frequencies of critical genes become different in the two groups. If these genetic differences are so great that the two groups can no longer interbreed to form fertile offspring, new species have been formed, i.e. **speciation** has occurred.

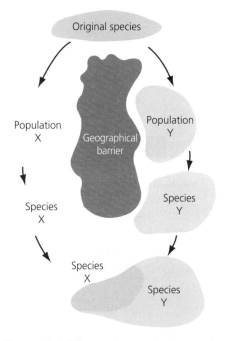

(1) Original species

(2) Physical barrier divides the species into two populations

(3) In different environments, different features have a selective advantage

(4) The two populations show increasing genetic differences

(5) The two populations are now distinct species; even after the barrier has gone they cannot interbreed

Figure 3.8 Allopatric speciation — the formation of new species when populations are geographically separated

Speciation is the formation of new species that can result from groups from one population becoming reproductively isolated and accumulating genetic differences.

Sympatric speciation

Reproductive isolation does not always involve physical separation. The two diverging groups might occupy the same area but be prevented from breeding in other ways:

- **seasonal isolation** — the two groups reproduce at different times of the year
- **temporal isolation** — the two groups reproduce at different times of the day
- **behavioural speciation** — members of the two groups have different courtship patterns

Speciation following any of these isolation mechanisms is called **sympatric speciation**.

An evolutionary race

REVISED

The first **antibiotic** was mass produced in the 1940s. Since then, the number of antibiotics has grown considerably, as has their use. As a result, many previously fatal diseases have been treated.

Unfortunately, natural selection works on bacterial populations too. A random mutation of a gene results in an allele conferring resistance to an antibiotic. In the absence of an antibiotic, this mutation will not give its possessor a competitive advantage over other bacteria; it will probably put it at a competitive disadvantage. In the presence of an antibiotic, however, susceptible bacteria will be killed and the one resistant bacterium will survive. By binary fission, this surviving bacterium will give rise to a new population, all members of which will carry the allele for resistance. At a stroke, evolution has occurred (Figure 3.9).

An **antibiotic** is a substance that disrupts part of the metabolism of bacterial cells.

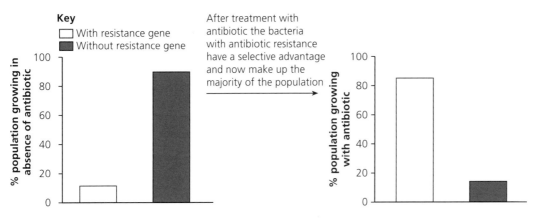

Figure 3.9 Directional selection in bacteria has led to antibiotic-resistant populations

16 In the absence of antibiotic X, a mutation to a gene resulting in an allele conferring resistance to the antibiotic will probably put the bacterium at a selective disadvantage. Suggest why.

Answer on p. 199

As more and more species of bacteria become resistant to one or more antibiotics, new pharmaceutical products are needed to treat the diseases they cause. This can be regarded as an **evolutionary race** between the bacteria and the pharmaceutical industry.

Biodiversity

Biodiversity can be assessed at different levels of organisation:
● variation within a habitat — looking at the variety of different species
● variation within a species — looking at the variety of alleles of each gene

Variation within a habitat

REVISED

We can measure the diversity of species within a **community** in one of two ways:
● **Species richness** is a measure of the number of different species present in a community. This measure has its limitations. For example, if one species is represented by only one or two individuals, it will not contribute much to the biodiversity of this habitat.
● **Species diversity** takes into account the number of individuals of each species as well as the number of species within a community. An **index of diversity** (D) can be found using the following formula:

$$D = \frac{N(N-1)}{\Sigma n(n-1)}$$

In this formula, N is the number of organisms of all species, n is the number of organisms of each species and the symbol Σ means 'the sum of'.

Variation within a species

REVISED

Within a **population**, all the organisms are of the same species.
Consequently, they all carry the same genes in the same positions on the

> **Typical mistake**
>
> Students often write that the presence of antibiotic causes mutations that lead to resistance. This is completely wrong; mutations occur randomly, by chance.

> **Biodiversity** is the variety of organisms living in a habitat.

> A **community** is all the different populations living in the same habitat at the same time.

> A **population** is a group of organisms of the same species living in the same habitat at the same time.

same chromosomes. These genes form the **gene pool** of that population. However, there might be several alleles of each gene within that gene pool. This gives rise to biodiversity within the gene pool.

> A **gene pool** is the sum of all the genes within a population, usually considered gene by gene. For example, in a population of 1000 diploid organisms, the gene pool for a particular gene will comprise 2000 copies of the gene.

Now test yourself

TESTED ☐

17 How would you expect the biodiversity of the gene pool of a large sexually reproducing population to differ from that of a very small sexually reproducing population? Explain your answer.

Answer on p. 199

The maintenance of biodiversity

REVISED ☐

The current human population is over 7 billion (7×10^9) people; it has doubled in the past 50 years. Such phenomenal population growth has been possible through the exploitation of natural resources, including:

- clearing existing ecosystems and replacing them with farmland
- using resources from ecosystems for construction, e.g. felling trees for timber
- using fossil fuels, leading to pollution and global warming

As a result, many ecosystems have been changed or destroyed. This results in the loss of organisms that inhabit those ecosystems and, consequently, a loss of biodiversity on a global scale.

Conservation involves managing the environment to maintain biodiversity. The aims of conservationists include:

- preserving and promoting existing habitats
- ensuring that natural resources are used in a way that encourages a sustainable yield

Some people regard conservation as a luxury for people in the developed world that denies people in the developing world the economic progress they desire. Table 3.5 summarises some economic and ethical reasons for promoting conservation.

> **Exam tip**
>
> Examiners expect you to show more understanding of biodiversity and conservation than general members of the public gain from newspaper articles.

> **Conservation** is the management of the environment to maintain biodiversity.

Table 3.5 Some economic and ethical reasons for promoting conservation

Type of reasoning	Notes

Economic	Natural ecosystems provide:
	● goods, such as timber, sources of medicines and fish
	● services, such as:
	  ○ tree roots help to stabilise soil
	  ○ photosynthetic organisms remove carbon dioxide from the environment and release oxygen into it
	● resources, such as a gene pool resource. Alleles of genes found in wild cereals, for example, might be useful in producing new crops that can withstand the effects of global warming

Ethical	We have a responsibility to maintain species, habitats and ecosystems for future generations
	Natural ecosystems are healthy areas for recreation and leisure
	Many ecosystems under threat are the home of indigenous peoples who have a right to their traditional way of life
	Many threatened habitats are home to other hominids. The rights of these hominids have recently become a worldwide concern

Ex-situ and in-situ conservation

Two approaches to conservation have emerged, *ex-situ* and *in-situ*. Some of the issues arising from the use of each method are summarised in Table 3.6.

> *Ex-situ* **conservation** is an attempt to preserve biodiversity by creating seed banks, botanical gardens and zoological gardens.
>
> *In-situ* **conservation** is an attempt to maintain biodiversity by designating and preserving representative habitats as nature reserves.

Table 3.6 Issues surrounding *ex-situ* and *in-situ* conservation methods

Ex-situ conservation	*In-situ* conservation
Botanical and zoological gardens can help the conservation of individual, threatened species but they cannot replicate entire ecosystems	The creation of nature reserves prevents the total loss of rare, natural habitats
Seed banks are an efficient way of maintaining the genetic material of endangered plant species so that future repopulation is possible	A nature reserve preserves the entire community of that habitat
Zoological gardens were designed originally to allow people to view unfamiliar animals, with little regard for their welfare	The community of a nature reserve can be monitored for early signs of deterioration, allowing remedial steps to be taken
Modern captive breeding programmes can boost numbers of healthy, mature members of an endangered species before release back into the wild	The offspring of endangered species are nurtured in their natural habitat, where they learn survival skills from their parents and peers
Inter-zoo breeding programmes can overcome the problems of inbreeding within a single zoo, thus maintaining some genetic diversity of an endangered species	Nature reserves are ideal places for the release of endangered individuals that are the product of captive breeding programmes
Visits to botanical and zoological gardens are easily accessible ways of raising awareness of the effect of humans on the world's ecosystems	When carefully managed, visits to nature reserves maintain public awareness of the susceptibility of ecosystems and of the need to manage for sustainability

Exam practice

1 Tigers have the biological name *Panthera tigris*. The table shows the biological classification of tigers.

Taxon	Name
Kingdom	Animalia
	Chordata
	Mammalia
	Carnivora
	Felidae
Genus	
Species	

→

(a) Complete the table by filling in the missing taxa and the missing names. [3]

(b) Leopards have the biological name *Panthera pardus*. What does this tell you about the relationship between tigers and leopards? [2]

2 A student sampled the plant communities in two different areas. Her results are shown in the table.

Species of plant	Number of plants (*n*)	
	Area A	Area B
Bramble	3	3
Buttercup	7	4
Holly	8	3
Sedge	7	4
Woodrush	4	2
Yorkshire fog	6	12
Total (*N*)	35	28

(a) What is meant by the term 'plant community'? [2]

(b) Compare the species richness of area **A** and area **B**. [1]

(c) Use the following formula to calculate the index of diversity (*D*) for each area. [2]

$$D = \frac{N(N-1)}{\Sigma n(n-1)}$$

(d) Compare the indexes of diversity for area **A** and area **B**. [1]

3 The animals named in the diagram are types of marine mammal. The diagram shows the DNA base sequences of part of a gene that the different animals have in common.

Species	DNA base sequence in part of gene that these species have in common
Fin whale	TAAACCCCAATAGTCA–CAAAACAAGACTATTCGCCAGAGTACTACTAGCAAC
Humpback whale	TAAACCCTAATAGTCA–CAAAACAAGACTATTCGCCAGAGTACTACTAGCAAC
Sperm whale	TAAACCCAGGTAGTCA–TAAAACAAGACTATTCGCCAGAGTACTACTAGCAAC
Bottlenose dolphin	TAAACTTAAATAATCC–CAAAACAAGATTATTCGCCAGAGTACTATCGGCAAC
Harbour porpoise	TAAACCTAAATAGTCC–TAAAACAAGACTATTCGCCAGAGTACTATCGGCAAC

(a) Name the bases represented by the letters A, C, G and T. [2]

(b) What is the maximum number of amino acids that could be encoded by the base sequences shown? [1]

(c) Complete the table to show the number of differences in the base sequences of these marine mammals. Parts of the table have been completed for you. [2]

	Fin whale	Humpback whale	Sperm whale	Bottlenose dolphin	Harbour porpoise
Fin whale					
Humpback whale					
Sperm whale					
Bottlenose dolphin	9	9	11		
Harbour porpoise	7	7	7		

(d) What caused the difference between the base sequences of the fin whale and humpback whale? [2]

(e) Which of the mammals in the table is most closely related to the humpback whale?
Use evidence from the table to support your answer. [2]

4 Health professionals have predicted that microbial resistance to antibiotics could cause 10 million deaths a year worldwide by 2050.

(a) Describe how a population of antibiotic-resistant bacteria could develop. [4]
Research published in September 2015 mapped patterns of antibiotic use and microbial resistance to antibiotics worldwide. The research highlighted two issues underpinning the growing threat of antibiotic-resistant infection by bacteria. It is almost 30 years since the development of a new class of antibiotics. The rise in antibiotic resistance is being accelerated by the overuse of antibiotics in healthcare and agricultural industries.

(b) Suggest two reasons why a new class of antibiotics has not been developed for almost 30 years. [2]

(c) One way in which antibiotics are being overused in the healthcare industry is the prescription of antibiotics for common colds and influenza. Explain why the prescription of antibiotics for common colds and influenza is considered to be 'overuse'. [2]

(d) Suggest why antibiotics are used in the agricultural industry. [2]

5 The table gives information about the area of the Amazon rainforest in Brazil.

Year	Estimated area of rainforest/km²
1970	4 100 000
1977	3 955 870
1987	3 744 570
1990	3 692 020
1997	3 576 965
2001	3 505 932
2004	3 432 147
2006	3 400 254

(a) Suggest how the area of rainforest could be estimated. [2]

Exam tip

When a question uses the command word 'suggest', the answer might be beyond the wording of the specification. Examiners expect you to use your understanding to propose an explanation that is reasonable from the information in the question.

(b) Plot a graph to show the area of rainforest remaining each year, calculated as a percentage of the area in 1970. [5]

(c) Explain how the loss of rainforest reduces biodiversity. [3]

6 (a) Give two reasons why it is often difficult to assign organisms to any one species. [2]

(b) How can the results of DNA sequencing help scientists to assign organisms to different species? [3]

(c) Explain the importance of peer review in assigning organisms to a particular species. [3]

Answers and quick quiz 3 online

ONLINE

Summary

Classification

- The classification system consists of a hierarchy of non-overlapping groups called taxa. The main taxa are: domain, kingdom, phylum, class, order, family, genus and species.
- Two classification systems are in current use:
  - The five-kingdom model in which the five kingdoms are: Prokaryotae, Protoctista, Fungi, Plantae and Animalia.
  - The three-domain model in which the Prokaryotae are divided into two domains, the Archea and Bacteria, and the other four kingdoms are classed in the domain Eukarya.
- Organisms are named using the name of their genus and species — the binomial system.
- The term 'species' can be defined as a group of organisms with similar characteristics that interbreed to produce fertile offspring.
- Determining new species can be difficult if the ability to interbreed to produce fertile offspring cannot be tested.
- Data obtained by DNA sequencing can be used to distinguish between species and to determine the evolutionary relationships between closely related species.
- Peer review of research reports before their publication in scientific journals and scientific conferences helps to validate new evidence.

Natural selection

- Natural selection brings about genetic changes in populations when genetic variation results in differential reproductive success within a population.
- Natural selection can be directional or stabilising.
- Natural selection has resulted in the development of bacterial populations that are resistant to one or more antibiotics.
- If two or more groups from a population become reproductively isolated, natural selection can bring about speciation.
- Reproductive isolation may occur through geographical separation of groups, leading to allopatric speciation, or without geographical separation of groups, leading to sympatric speciation.

Biodiversity

- Biodiversity can be assessed at different levels:
  - within a habitat at the species level
  - within a species at the genetic level
- Within a community, an index of diversity (D) can be calculated from sampling data using the formula:

$$D = \frac{N(N-1)}{\Sigma n(n-1)}$$

  where N is the number of organisms of all species and n is the number of organisms of each species.
- The process of conservation is used to maintain biodiversity and the sustainable use of natural resources.
- The maintenance of biodiversity can be justified using economic and ethical reasons.
- *Ex-situ* conservation involves establishing seed banks, botanical gardens and zoological gardens to maintain biodiversity. *In-situ* conservation involves designating and protecting areas of natural habitat.

4 Exchange and transport

Organisms remain alive as a result of chemical reactions occurring within their cells. Collectively, we call these reactions **metabolism**. Metabolic reactions use reactants, many of which originate from outside the organism. They generate products, some of which are not useful and are **excreted** by the organism. All organisms gain essential reactants and excrete waste products across **exchange surfaces**.

> **Excretion** is the elimination by an organism of substances produced by its own metabolism.

Surface area to volume ratio

The ratio of surface area to volume is a recurring theme in biology. In terms of metabolism, it is related to supply and demand. **Supply** is the rate at which metabolites can enter a cell or organism and is dependent on the surface area of the relevant exchange surface. **Demand** is the rate at which an organism uses its metabolites and is dependent on the number of active cells in its body. The more the number of active cells, the greater the volume of the organism.

Figure 4.1 shows the relationship between surface to volume ratio and body size of organisms. You can see that the larger the organism, the smaller the surface area to volume ratio (and hence rate of supply of metabolites).

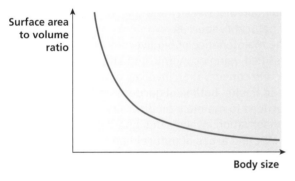

Figure 4.1 The relationship between body size and surface area to volume ratio

Large active organisms have adaptations that overcome the limitations of a small surface area to volume ratio. These include:
- adaptations that increase the area of exchange surfaces, e.g. extensions of the exchange surface outwards (fish gills) and extensions of the exchange surface inwards (mammalian lungs)
- mass transport systems — blood circulatory systems in animals; xylem and phloem in plants

> **Revision activity**
>
> Using the following formulae, calculate the surface area to volume ratio of six spheres of radius 1, 2, 3, 4, 5 and 6 cm. Plot the surface area to volume ratio of these spheres against their radius. Compare your graph with Figure 4.1.
>
> Surface area = $4\pi r^2$
>
> Volume = $\frac{4}{3}\pi r^3$

> **Exam tip**
>
> Don't imply that organisms evolve features *in order to be* better adapted to their habitat.

Cell transport mechanisms

The cell is the basic unit of metabolism. It is surrounded by a **cell surface membrane** through which all metabolites enter and all waste products leave.

The fluid-mosaic model of membrane structure

REVISED

Every cell surface membrane and every membrane within the cytoplasm of eukaryotic cells has the structure shown in Figure 4.2.

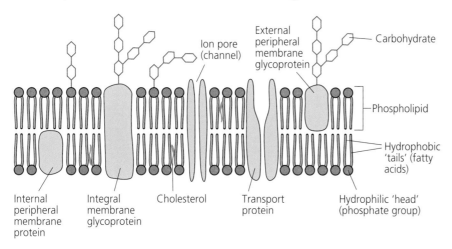

Figure 4.2 The fluid-mosaic model of membrane structure

The membrane is **fluid** because the molecules within it can move around laterally. The membrane is a **mosaic** because it contains many different types of molecule. These include:

- two layers of phospholipid molecules (a **phospholipid bilayer**) — the hydrophilic 'heads' are on the outside of both layers, in contact with water-containing fluids; the hydrophobic 'tails' are on the inside of the membrane
- **protein** molecules embedded within the phospholipid bilayer, some of which have attached oligosaccharide chains (making them **glycoproteins**).
    - proteins present only on the outside of the membrane (external peripheral membrane proteins) are often specific receptor molecules
    - proteins present only on the inside of the membrane (internal peripheral membrane proteins) often help to anchor other proteins in position in the cytoplasm
    - proteins that pass through the membrane (transmembrane, or integral membrane proteins) act as pores, channels and carriers for particles that cannot freely cross the phospholipid bilayer
- **cholesterol** molecules embedded within the phospholipid bilayer prevent the membrane becoming too fluid

Now test yourself

TESTED

1 What are the hydrophobic 'tails' of a phospholipid made from?

Answer on p. 199

The movement of substances across cell surface membranes

REVISED

The phospholipid bilayer does not allow the free movement of particles through it. For this reason, the cell surface membrane is referred to as **partially permeable**.

Passive transport

Passive transport relies on the random thermal movement of particles that occurs in all solids, liquids and gases. Consequently, passive transport across cell surface membranes does not involve any energy expenditure by the cell. Table 4.1 summarises what you need to understand about the three types of passive transport that occur across cell surface membranes.

Table 4.1 **Three types of passive transport across cell surface membranes**

Type of passive transport	Type of particle that can cross cell surface membrane	Further details
Diffusion	Small, uncharged molecules Lipid-soluble molecules	Diffusion is the free movement of particles from a region of their high concentration to a region of their low concentration (i.e. down a concentration gradient) until the concentrations are equal
Facilitated diffusion	Charged particles Larger non-lipid-soluble molecules	The movement of particles down a concentration gradient either: ● through protein molecules that form **ion channels**, e.g. sodium ions, or ● carried by **transport proteins**, e.g. glucose and amino acids
Osmosis	Water molecules	The movement of water molecules across a partially permeable membrane from a high **water potential** to a lower (more negative) water potential Protein channels, called **aquaporins**, increase the ability of water molecules to pass through the phospholipid bilayer

Water potential (Ψ) is a measure of the concentration of free water molecules in a solution. Pure water has a water potential of 0 kPa. Solutions have lower (more negative) values of water potential than that of water, e.g. −6 kPa.

Exam tip

Be sure to use the term 'water potential' when explaining osmosis and to make clear that *water moves down a water potential gradient* (which you could also express as from a higher to a lower water potential).

Now test yourself

TESTED

2 In addition to the surface area to volume ratio of an exchange surface, the rate of diffusion is affected by the thickness of the exchange surface and the concentration gradient across it. In each case, explain why.
3 A cell with a water potential of −10 kPa is placed in a solution with a water potential of −5 kPa. Will the cell gain water or lose water? Explain your answer.

Answers on p. 199

Active transport

Active transport differs from passive transport in three ways:

● It moves particles from a low concentration to a higher concentration (i.e. against, or up, a concentration gradient).
● It involves the hydrolysis of ATP by the cell (hence it is active transport).
● It always involves the activity of specific carrier proteins (often called **pumps**).

The biological role of ATP

A molecule of ATP (adenosine triphosphate) is similar to a nucleotide. It contains a deoxyribose residue, an adenine base and three phosphate groups (Figure 4.3). ATP is formed during the cellular process of **respiration**, during which:

● an inorganic phosphate group (denoted P_i) is added to a molecule of adenosine diphosphate (ADP)
● the energy required for this process results from the stepwise breakdown of a respiratory substrate

The hydrolysis of ATP to ADP and P_i releases energy.

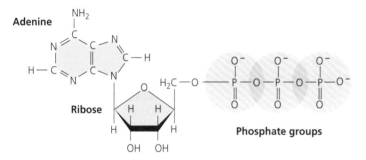

Figure 4.3 A molecule of adenosine triphosphate (ATP)

All cells link energy-driven reactions to the hydrolysis of ATP (Figure 4.4). For this reason, ATP is often described as the **universal energy currency** in cells. The release of energy by the hydrolysis of ATP is:

● instantaneous — energy is available whenever it is needed
● sufficient to drive the linked reaction but not to cause damaging heat gains in the cell

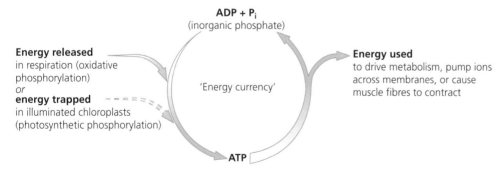

Figure 4.4 The hydrolysis of ATP is linked to energy-driven reactions

Now test yourself

TESTED

4 Every day, you make a mass of ATP that is greater than the mass of your body. Explain why this does not lead to a gain in your body mass.

Answer on p. 199

> **Exam tip**
>
> Avoid writing that particles move along a concentration gradient. To gain credit in an exam, you must specify the direction of movement, i.e. down or against a concentration gradient.

> Respiration is the cellular processes that produces ATP from ADP and P_i.

> **Exam tip**
>
> Be careful not to confuse respiration (the production of ATP) with breathing and gas exchange.

Endocytosis and exocytosis

Many eukaryotic cells pass substances into and out of their cytoplasm using membrane-bound vesicles.

During **endocytosis**, material outside the cell is brought into the cell. The cell extensions engulf the extracellular material. As the arms of the cell extensions touch, their membranes merge, forming a vacuole within the cytoplasm that contains the extracellular material.

During **exocytosis**, material inside the cell is released outside the cell. Membrane-bound vacuoles containing this material bud off from the flattened membranous sacs of the Golgi apparatus. These vesicles move through the cytoplasm to the cell surface membrane. The membrane of the vesicle merges with the cell surface membrane and the contents of the vesicle are released.

Gas exchange

Gas exchange involves the exchange of carbon dioxide and oxygen between organisms and their environment.

Aerobic respiration uses oxygen and produces carbon dioxide. The uptake of oxygen and release of carbon dioxide occurs in all aerobic organisms all of the time.

Photosynthesis uses carbon dioxide and produces oxygen. The uptake of carbon dioxide and release of oxygen occurs in plants whenever the intensity of light is high enough to allow photosynthesis.

> **Typical mistake**
>
> Students often fail to realise that plants respire all the time and not just at night.

TESTED

Now test yourself

5 During bright sunlight, a plant releases oxygen and takes in carbon dioxide. What does this tell you about the rate of respiration in the plant?

6 Suggest why a single-celled organism is able to survive without a specialised gas exchange surface.

Answers on p. 199

Multicellular organisms show a variety of adaptations that allow efficient gas exchange. You need to be familiar with four: insects, fish, mammals and flowering plants.

> **Exam tip**
>
> If you are asked to explain how a given system is adapted for efficient gas exchange, make sure your answer covers: large surface area to volume ratio; thinness of exchange surface; maintenance of diffusion gradient.

Gas exchange in insects

REVISED

Gas exchange in insects occurs by diffusion directly between the atmosphere and the insect's body cells. Figure 4.5(a) shows an insect's gas exchange system. Two main **tracheae** run the length of the insect's body. **Spiracles** along the length of the insect's body open from the atmosphere into these tracheae. The tracheae branch into smaller **tracheoles** that carry air directly to the body cells.

The large number of tracheoles creates a large surface area and their thin walls ensure a short diffusion distance between cells and the atmosphere. Both these features ensure efficient gas exchange.

Although gas exchange occurs by diffusion in most insects, some larger insects are able to speed this process. By opening some spiracles and closing others at the same time as dilating and constricting their abdomens, these insects can move air along the tracheae.

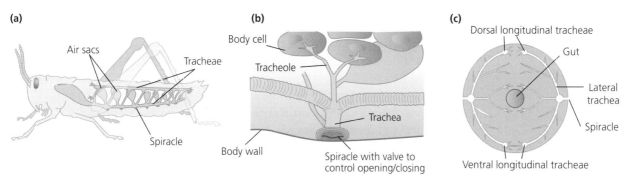

Figure 4.5 (a) The tracheal system of an insect (b) The relationship between spiracles, tracheae and tracheoles (c) Transverse section of the tracheal system

Now test yourself

TESTED ☐

7 When insects are not active, many close their spiracles. Suggest the biological advantage of this.
8 Most insects are small. Suggest why their gas exchange system severely limits the maximum size to which insects can grow.

Answers on p. 199–200

Gas exchange in fish

REVISED ☐

The gas exchange surface of a fish is its gills. Figure 4.6 shows the gills of a bony fish. Each gill:
● is attached to a bony gill arch
● is divided into a large number of gill filaments, which increase its surface area and have many gill lamellae (singular: lamella) that increase the surface area of the gill still further

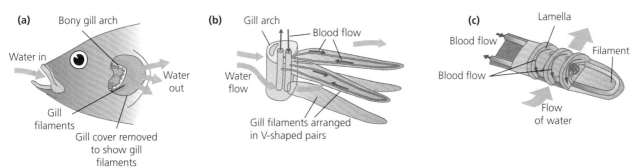

Figure 4.6 (a) The gills of a bony fish (b) The structure of the gill filaments (c) The countercurrent of blood flow in the lamellae and water flow over the lamellae

In addition to providing a large gas exchange surface, the gill lamellae have very thin walls, which result in a short diffusion pathway.

Unlike insects, fish have a blood circulatory system. This further increases the rate of gas exchange in two ways:

- Blood flow maintains a steep concentration gradient of gases between the surrounding water and the blood. As oxygen diffuses from the water into the blood, the blood with a high oxygen concentration is carried away to the tissues and replaced in the gill lamellae by blood with a low oxygen concentration. As carbon dioxide diffuses from the blood into the surrounding water, the blood with a low carbon dioxide concentration is replaced in the gill lamellae by blood with a high carbon dioxide concentration.
- As Figure 4.6(c) shows, blood within the lamellae flows in the opposite direction from the flow of water over the gills. This **countercurrent** further ensures that steep concentration gradients of oxygen and carbon dioxide are maintained between water and blood.

Some species of fish simply swim with their mouths open, allowing water to flow over their gills. However, many ventilate their gills by lowering the floor of their buccal cavity to draw water into the mouth, and closing the mouth and raising the floor of their buccal cavity to push water out over the gills.

Now test yourself

TESTED

9 Explain why the countercurrent flow of water and blood allows faster diffusion of oxygen and carbon dioxide across the gills of a fish.
10 Gills form the gas exchange surface of many groups of aquatic animals. Suggest why gills would not provide an efficient gas exchange system for terrestrial animals.

Answers on p. 200

Gas exchange in mammals

REVISED

The gas exchange surface of a mammal is the alveoli within its lungs. Figure 4.7 shows the gas exchange system of a human, one type of mammal. Its key features are:

- a single **trachea**, which branches into
- two **bronchi** (singular bronchus), one travelling to each lung, which divide into
- smaller and smaller **bronchioles**
- the smallest of which end in a group of **alveoli**

Although the lungs occupy a relatively small volume, those of an adult human have so many alveoli that their combined surface area has been estimated to be about the size of a doubles tennis court. You can see in Figure 4.7 that the alveoli are in direct contact with a capillary network. Figure 4.8 shows that the:

- wall of an alveolus is a single layer of cells thick
- wall of a capillary is a single layer of cells thick
- cells forming the walls of alveoli and capillaries are very thin cells, called **squamous cells**

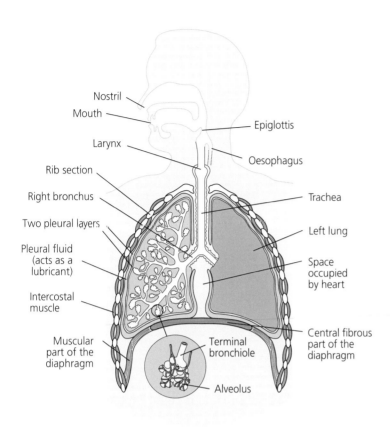

Figure 4.7 The gross structure of the human gas exchange system

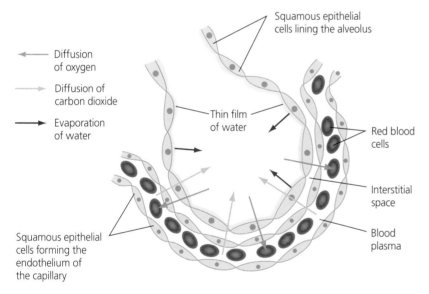

Diffusion of oxygen

Diffusion of carbon dioxide

Evaporation of water

Squamous epithelial cells lining the alveolus

Thin film of water

Red blood cells

Interstitial space

Blood plasma

Squamous epithelial cells forming the endothelium of the capillary

Figure 4.8 Gas exchange between the air in an alveolus and blood in the surrounding capillary

Just as many species of fish ventilate their gills, mammals ventilate their lungs. In so doing, they inhale air with a high oxygen concentration and a low carbon dioxide concentration, and exhale air with a low oxygen concentration and a high carbon dioxide concentration.

The lungs are located within an 'airtight' compartment — the **thorax** — surrounded by pairs of **ribs** and a muscular **diaphragm**. **Antagonistic pairs** of muscles, the internal and external **intercostal muscles** (intercostal means 'between the ribs'), are attached to each rib. Contraction of the internal intercostal muscles pulls the ribs downwards and inwards; contraction of the external intercostal muscles pulls the ribs upwards and outwards.

Exam tip

The film of water lining each alveolus is an unavoidable source of water loss. Don't report it as an adaptation for efficient gas exchange.

Revision activity

Draw two flow charts, one tracing the diffusion pathway of oxygen in the alveoli and a second tracing the diffusion pathway of carbon dioxide in the alveoli.

Antagonistic pairs are muscles that work in pairs in which one contracts as the other relaxes. In this way, one pair brings about two different effects on the skeleton.

Mammals ventilate their lungs by breathing. Figure 4.9 summarises how inhalation (breathing in) is brought about during normal 'quiet' breathing.

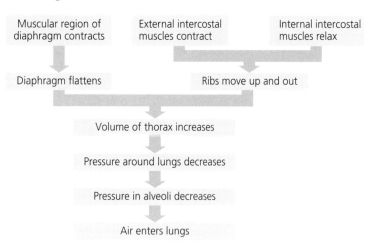

Figure 4.9 The mechanism of inhalation (breathing in)

Exam tip

Ensure you make clear that air moves into and out of the lungs by being pushed down a pressure gradient.

Now test yourself

TESTED

11 The lining of the alveoli and of the capillaries is a single layer of squamous cells. Explain how this is an adaptation that allows efficient gas exchange.
12 Draw a flow chart, like the one in Figure 4.9, to represent the mechanism of exhalation (breathing out).
13 Suggest when contraction of the internal intercostal muscles and relaxation of the external intercostal muscles might occur.

Answers on p. 200

Gas exchange in flowering plants

REVISED

Gas exchange in a flowering plant occurs in two areas: in the **stomata** in the leaves, and in the **lenticels** in the stem. In each case, gas exchange is by diffusion between living cells and the surrounding atmosphere, with no ventilation mechanisms.

In the leaves

Figure 4.10 shows a vertical section through a leaf. It has:
- **stomata** that are open to the air. Most of the stomata are on the lower leaf surface
- cells in the **spongy mesophyll** region that are so loosely packed that they provide a surface area that is large enough for efficient gas exchange
- **air spaces** within the spongy mesophyll that create a diffusion pathway from the stomata to the palisade cells at the top of the leaf

Terrestrial organisms always face water loss by evaporation from their permeable surfaces. Just as many species of insect can limit water loss by closing their spiracles, many plant species can limit water loss by closing their stomata.

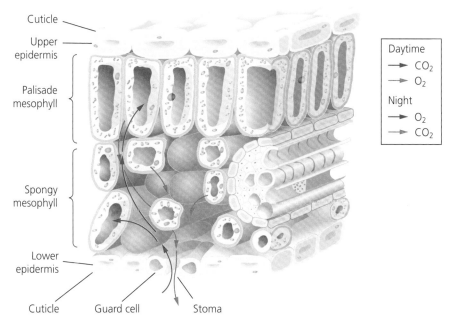

Cuticle

Upper epidermis

Palisade mesophyll

Spongy mesophyll

Lower epidermis

Cuticle   Guard cell   Stoma

Daytime
→ CO₂
→ O₂
Night
→ O₂
→ CO₂

Figure 4.10 Net gas exchange in a leaf during the day and at night

You can see in Figure 4.10 that each stoma (plural: stomata) has two **guard cells** around it. These guard cells are sausage-shaped. When water diffuses by osmosis into the cytoplasm of these cells, they expand in such a way that they open the stoma they surround.

In the stems

The stem of many plants have openings called **lenticels**. Each is open to the atmosphere and has loosely packed cells behind it. Just as with the leaf, the loosely packed cells provide:

- a large surface area for gas exchange
- a diffusion pathway to the living cells in the outer layer of the stem. (In woody stems, most of the cells forming the trunk are dead xylem cells. Only the outer cells are living.)

Circulation

Over large distances, diffusion becomes an inefficient transport method. Many multicellular animals can remain active because they have evolved circulatory systems that enable efficient mass transport.

Types of animal circulatory system

REVISED

Animal circulatory systems transport substances in a fluid called blood. The systems may be open or closed.

Open circulatory systems lack blood vessels. The blood of an insect is housed within a body cavity, called a **haemocoel**. Contractions of a muscular tube within this cavity mix up the blood.

Closed circulatory systems have a heart that pumps blood around the system within blood vessels. Closed systems can be single or double:

- A **single circulatory system** is found in fish. The flow of blood is:

  heart → capillaries in gills → capillaries in body tissues → heart

  The heart of a fish is a single pump, with one receiving chamber (the **atrium**) and one pumping chamber (the **ventricle**).

> **Exam tip**
>
> Don't write that animals have evolved circulatory systems *in order to* transport substances efficiently. This implies they had the level of understanding you have and consciously developed them.

- A **double circulatory system** is found in mammals. The flow of blood in a double system is:

heart → capillaries in lungs → heart → capillaries in body tissues → heart

The heart of a mammal has two pumps, each containing an atrium and a ventricle. One pumps deoxygenated blood to the lungs and the other pumps oxygenated blood to the body tissues.

The advantages of a double circulatory system

In both single and double circulatory systems, contraction of the heart increases the **hydrostatic pressure** in the blood, which pushes the blood through the blood vessels. As blood flows away from the heart, its hydrostatic pressure falls. This fall in pressure is particularly great in the smaller arterioles and capillaries of the gas exchange surface and the body tissues.

In a single circulatory system, this fall in blood pressure occurs twice for each circulation of blood through the system: once in the capillaries of the gills and once again in the capillaries of the body tissues.

In a double circulatory system, this fall in blood pressure occurs once in the circulation to the lungs and once in the circulation to the body tissues. Between each, blood return to the heart increases the blood pressure. Consequently, a double circulation allows more rapid circulation of blood than a single circulation system.

> **Hydrostatic pressure** refers to the pressure that any fluid in a confined space exerts. If fluid is in a container, there will be some pressure on the wall of that container.

Now test yourself

TESTED

14 (a) Why does blood pressure fall as blood is pushed through blood vessels?
(b) Suggest why this fall in blood pressure is particularly great in smaller arterioles and capillaries.

Answers on p. 200

The mammalian circulatory system

REVISED

Figure 4.11 shows the main components of the mammalian circulatory system. Deoxygenated blood is shown in blue and oxygenated blood is shown in red.

The heart

The **heart** has four chambers:
- The **right atrium** receives deoxygenated blood from the body and the **right ventricle** pumps this deoxygenated blood to the lungs.
- The **left atrium** receives oxygenated blood from the lungs and the **left ventricle** pumps this oxygenated blood to the body tissues.

Blood vessels

Arteries carry blood at high pressure away from the heart to the body tissues. As they do so, they branch into smaller and smaller **arterioles**.

Veins carry blood under low pressure from the body tissues back to the heart. Veins are formed by the fusion of smaller vessels called **venules**.

Capillaries are vessels that carry blood from arterioles to the surrounding body tissues and then to venules.

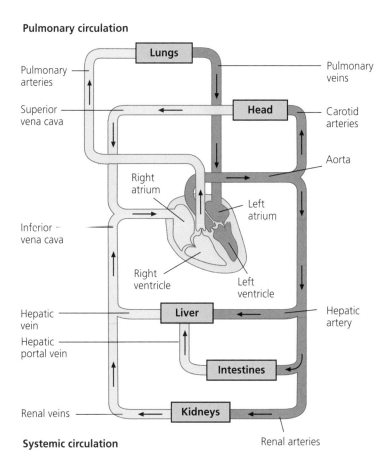

Figure 4.11 The main components of the mammalian circulatory system

Figure 4.12 shows the structure of an artery, a vein and a capillary. Each has a lining comprising a single layer of squamous cells. In addition, an artery and vein have elastic tissue, muscle tissue and a protective outer layer.

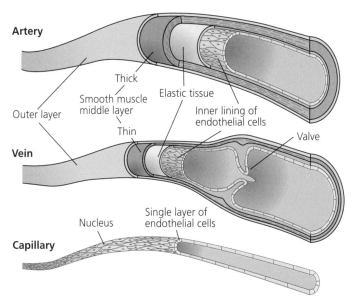

Figure 4.12 The structure of an artery, a vein and a capillary

Arteries

Arteries are adapted to their function in the following ways:

- The layer of elastic tissue in an artery is thicker than that in a vein. This enables arteries close to the heart to overcome the pulsatile nature of blood flow from the ventricles. The elastic tissue allows the larger arteries to expand, accommodating the blood from the ventricles, and then recoil, smoothing out the flow of blood.
- The layer of muscle is thicker in an artery than in a vein. This enables the arterioles near the tissues to dilate (**vasodilation**) and constrict (**vasoconstriction**). As a result, mammals are able to control the flow of blood to the different organs of the body.

> **Exam tip**
>
> Always make it clear that it is *arterioles* that are involved in vasoconstriction and vasodilation.

Veins

Veins are adapted to their function in the following ways:

- Veins possess semilunar valves along their length. As the blood pressure in the veins is low, there is a danger that blood will either remain stationary or move backwards. Any backflow of blood will fill the 'pockets' of the semilunar valves, pushing them together so that the valves close the lumen of the vessel.
- The thinner walls of veins allow contraction of the skeletal muscles around the veins to prevent the blood becoming stationary and to push blood back to the heart.

Capillaries

Capillaries are adapted to their function in the following ways:

- Capillaries have a very small diameter. This greatly increases their surface area to volume ratio, allowing rapid exchange across their surface.
- The walls of capillaries consist of a single layer of squamous cell. This reduces the distance for exchange across their surface.

Now test yourself

TESTED

15 Return of blood to the heart of a musician playing a long note on a wind instrument could stop momentarily. Suggest why.

Answer on p. 200

> **Revision activity**
>
> Write a brief explanation of how arteries are adapted for their function. As this is an explanation, you should give a reason why each feature you describe adapts arteries for their function.

The heart and cardiac cycle

REVISED

The structure of the heart

You should be able to identify the parts of the heart shown in Figure 4.13. The left and right sides are separated by a muscular wall, the **septum**. On each side, the upper atrium is separated from the lower ventricle by an **atrioventricular valve**. Between each ventricle and the artery leaving it is a **semilunar valve**.

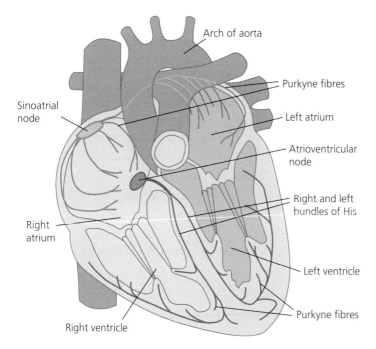

Figure 4.13 **A vertical section through the heart of a mammal**

The cardiac cycle

The duration of a single heartbeat is called the **cardiac cycle**. The events occurring in a single cycle are shown in Figure 4.14 and summarised in Table 4.2. Notice that the events in one atrium are mirrored by those in the other, as are the events in the ventricles.

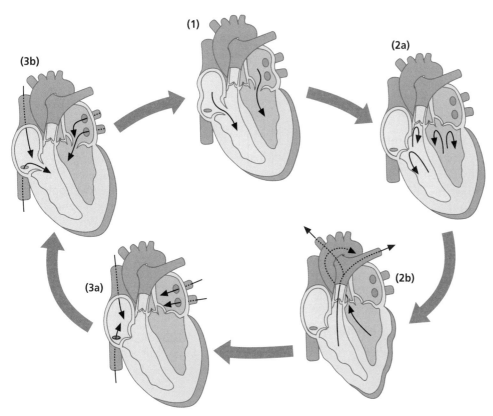

Figure 4.14 **The events occurring during one cardiac cycle**

Table 4.2 An explanation of the events occurring during one cardiac cycle

Stage from Figure 4.14	Atria		Ventricles	
	Action of muscles in walls	Result	Action of muscles in walls	Result
1 Atrial systole	Contract	Increase in pressure pushes open the atrioventricular valves Blood is pushed into the ventricles	Relax	Ventricles expand as they fill with blood
2 Ventricular systole	Relax	Blood neither enters nor leaves	Contract	Pressure of blood increases Eventually, the blood pressure is strong enough to push the atrioventricular valves closed and the pulmonary and aortic valves (the semilunar valves) open. Blood is then pushed into the arteries
3 Diastole	Relax	Blood flows into atria As blood pressure increases, it pushes the atrioventricular valves open	Relax	Back pressure of blood in the arteries closes the aortic and pulmonary valves. Blood neither leaves nor enters the ventricles Blood then enters the ventricles as a result of the higher pressure in the atria

Control of the cardiac cycle

Cardiac muscle is **myogenic**. Specialised muscle cells, shown in Figure 4.13, produce regular electrical impulses that start the cardiac cycle and transmit impulses from cell to cell throughout the heart. The **sinoatrial node** (**SAN**) is a bundle of cells in the wall of the right atrium that generates electrical impulses. These electrical impulses spread through the walls of both atria and stimulate atrial systole. The impulses can only pass through to the walls of the ventricles through the **atrioventricular node** (**AVN**). The impulses are delayed here, causing the delay between atrial systole and ventricular systole. From the AVN, the impulses pass down the septum in specialised muscle cells forming the **bundle of His**. From the base of the bundle of His, specialised **Purkyne fibres** carry impulses through the walls of the ventricle from the base of the heart upwards. This causes each ventricle to contract from its base upwards.

The electrical changes that occur in the heart tissue during a single cardiac cycle can be measured and displayed as an electrocardiogram (**ECG**), as shown in Figure 4.15.

> **Exam tip**
>
> When describing the opening and closing of valves during the cardiac cycle, think of two people pushing against a door from opposite sides. Whoever pushes more strongly moves the door.

> **Myogenic** is the ability of cardiac muscle to contract without stimulation by the nervous system.

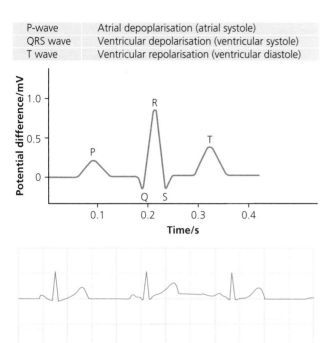

P-wave	Atrial depolarisation (atrial systole)
QRS wave	Ventricular depolarisation (ventricular systole)
T wave	Ventricular repolarisation (ventricular diastole)

A complete ECG trace from a healthy patient

Figure 4.15 An electrocardiogram shows the electrical changes during one cardiac cycle

Exam tip

The abbreviations AVN, SAN and ECG are given in the specification, so you can use them in your answers to exam questions without further elaboration.

Now test yourself

TESTED

16 Impulses carried by the Purkyne fibres cause the ventricles to contract from their bases upwards. Suggest the biological advantage of this.

Answer on p. 200

Mammalian blood

REVISED

The function of mammalian blood can be summarised as:
- **transport** — all metabolites and waste products are carried within the blood
- **formation of tissue fluid** — the fluid that surrounds cells and in which metabolites are exchanged between the blood and the tissues
- **defence** against foreign bodies

Blood composition

Blood is made up of:
- **plasma** — the fluid comprising water, dissolved molecules and ions, and suspended proteins
- **erythrocytes** (red blood cells) — these transport oxygen in the form of oxyhaemoglobin and play a role in the transport of carbon dioxide as hydrogencarbonate ions (HCO_3^-)
- **leucocytes** (white blood cells) — these play a role in defence mechanisms:
  - **lymphocytes** release antibodies (B lymphocytes) or chemicals that aid the immune response (T lymphocytes)
  - phagocytes such as **monocytes** and **neutrophils** ingest and destroy bacteria or harmful chemicals, respectively
  - **eosinophils** contain enzymes that detoxify foreign proteins
- **platelets** — cell fragments that initiate blood clotting

Exam tip

You may refer to red blood cells and white blood cells in an examination, but you must use the technical names lymphocytes, monocytes, neutrophils and eosinophils where relevant.

The role of platelets and plasma proteins in blood clotting

If blood vessels are damaged, such as by a cut in the skin or a small internal haemorrhage, blood clots perform two functions:
● to reduce or stop blood loss
● to prevent the entry of **pathogens**

A wound triggers the **cascade of events** shown in Figure 4.16. The result is a **blood clot**.

> **Pathogens** are organisms that cause harm by damaging invading cells or releasing toxins.
>
> A **blood clot** is a network of insoluble protein fibres, fibrin and trapped red blood cells.

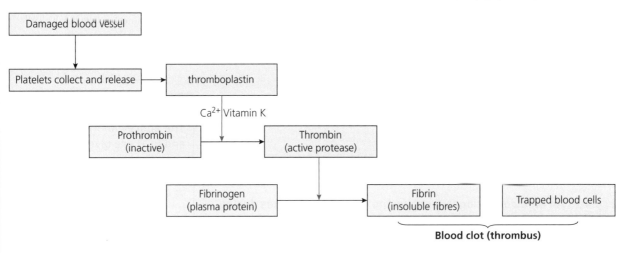

Figure 4.16 Blood clot formation

TESTED

Now test yourself

17 The containers used to store blood for transfusion contain a solution that removes calcium ions. Suggest the advantage of this.

Answer on p. 200

Atherosclerosis and myocardial infarction

REVISED

Blood clots are beneficial when they prevent blood loss and the entry of pathogens. They can be harmful when they form within a blood vessel.

A blood clot that forms within a blood vessel is called a **thrombus**. If a thrombus becomes dislodged, it can be carried away in the blood. It is now known as an **embolus**. An **embolism** occurs if an embolus blocks a small arteriole, stopping the blood supply to the downstream tissue. Without a supply of metabolites, such as glucose and oxygen, that tissue dies. If this happens in one of the branches of a coronary artery, the downstream heart tissue dies. This is known as a **myocardial infarction** or a heart attack, which can be fatal. If this happens in an arteriole in the brain, the downstream brain tissue dies. This is called an ischaemic **stroke** (contrasting with a haemorrhagic stroke, when a weakened arteriole in the brain bursts).

> **Typical mistake**
>
> Students often write that, following a myocardial infarction, the heart dies, rather than that *muscle tissue/muscle cells* supplied by the blocked arteriole die.

The most common cause of a myocardial infarction is **atherosclerosis** — the progressive degeneration of artery walls. Healthy artery walls have a smooth squamous epithelium lining. Atherosclerosis results as follows:

1 Deposits of lipoprotein and cholesterol, called **atherosclerotic plaque**, develop within the artery wall.

2 Atherosclerotic plaque breaks the lining of the arteriole, resulting in a roughened surface.

3 Platelets accumulate at the roughened surface and release thromboplastin, starting the blood–clotting process.

Research has linked a large number of factors with an increased risk of atherosclerosis. These include factors over which we have no control and those that involve lifestyle choices. Table 4.3 lists some of these risk factors.

Exam tip

Be sure to point out that an atherosclerotic plaque develops *within* the artery wall. You will not gain credit if you write that it occurs inside the artery as this could mean within the lumen of the artery.

Table 4.3 Risk factors associated with atherosclerosis and myocardial infarction

Risk factor		Notes
Factors over which we have no control	Age	The risk is greater with increasing age
	Genes	Atherosclerosis is known to run in families, is more common in males than in females and is more common in some ethnic groups than in others
Factors that involve lifestyle choices	Smoking tobacco	Risk is higher in smokers than in non-smokers
	Excessive alcohol consumption	Alcohol consumption leads to higher concentrations of LDL in the blood (see LDL below)
	Lack of physical activity	Even moderate physical activity reduces the risk of atherosclerosis
	Obesity	Obesity is associated with an increased risk of atherosclerosis
Lifestyle choices and/ or genetics	High concentration of low-density lipoprotein (LDL) in the blood	High LDL concentrations lead to increased risk of atherosclerosis A high concentration of LDL could be related to diet, obesity and diabetes, but is also more common in some ethnic groups than in others, irrespective of these other factors

Transport of gases in the blood

The transport of oxygen from a gas exchange surface to actively respiring cells and of carbon dioxide in the opposite direction maintains steep concentration gradients at the gas exchange surface. Red blood cells play a vital role in the transport of both gases.

Oxygen

REVISED

Each of the four haem groups in a molecule of haemoglobin (represented as Hb) can combine reversibly with one molecule of oxygen to form oxyhaemoglobin (represented as HbO_8).

$$4O_2 + Hb \rightleftharpoons HbO_8$$

Carbon dioxide

REVISED

Carbon dioxide transport occurs in two ways:
- As **hydrogencarbonate ions** in the red blood cells and in the plasma. An enzyme, **carbonic anhydrase**, within red blood cells catalyses the reaction:

$$CO_2 + H_2O \rightleftharpoons HCO_3^- + H^+$$

- As **carbaminohaemoglobin** in red blood cells, formed by a reaction between carbon dioxide and the amino groups within the polypeptide chains of haemoglobin:

$$CO_2 + Hb \rightarrow carbaminohaemoglobin + H^+$$

Now test yourself

TESTED

18 Suggest one advantage of haemoglobin being packaged within red blood cells rather than being in suspension in the plasma.

Answer on p. 200

The oxygen dissociation curve of haemoglobin

REVISED

Reading the *x*-axis of Figure 4.17 from left to right shows how the percentage saturation of haemoglobin increases as the partial pressure of oxygen increases.

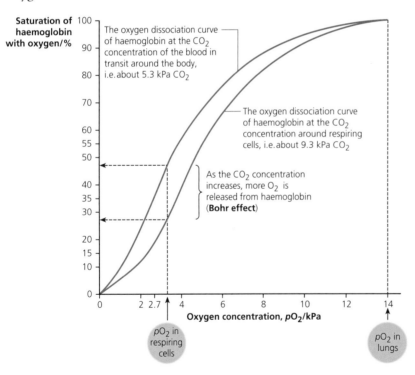

Figure 4.17 **The oxygen dissociation curve**

The curve has a distinctive S-shape because the combination of a molecule of haemoglobin with the first oxygen molecule makes it easier to add the next two oxygen molecules, but the final oxygen molecule is more difficult to add.

Reading the x-axis of Figure 4.17 from right to left represents the breakdown of oxyhaemoglobin to release oxygen (i.e. **oxyhaemoglobin dissociation**) as the partial pressure of oxygen falls.

At a high partial pressure of oxygen in the lungs, haemoglobin is fully saturated. At lower partial pressures of oxygen in the respiring tissues, oxyhaemoglobin dissociates, releasing oxygen. Higher carbon dioxide concentrations in the blood cause oxyhaemoglobin to dissociate more readily, shifting the dissociation curve along the x-axis to the right (see Figure 4.17). This is called the **Bohr effect**.

Fetal haemoglobin has a structure that is different from adult haemoglobin, which results in fetal haemoglobin combining more readily with oxygen than does adult haemoglobin. In a graph, the dissociation curve for fetal haemoglobin is to the left of that for adult haemoglobin.

Myoglobin is an oxygen-carrying molecule found in muscle tissue. A molecule of myoglobin is a conjugated protein with only one haem group. As a result, myoglobin combines more readily with oxygen than does haemoglobin. In a graph, the dissociation curve for myoglobin is to the left of that for adult haemoglobin.

> **Oxyhaemoglobin dissociation** is the breakdown of oxyhaemoglobin to release the oxygen it carries.

> **Exam tip**
>
> To understand and explain the oxygen dissociation curve of haemoglobin, you must read the x-axis from right to left.

> The **Bohr effect** occurs when oxyhaemoglobin dissociates more readily at high carbon dioxide concentrations than at lower carbon dioxide concentrations.

Now test yourself

TESTED

19 Explain the advantage of the Bohr effect in the active muscle cells of an adult human.
20 On a graph showing oxygen dissociation, the curves for fetal haemoglobin and myoglobin would be to the left of that for adult haemoglobin. Explain the advantage of the positions of the curves for fetal haemoglobin and myoglobin.

Answers on p. 200

Transfer of materials between the circulatory system and cells

The formation and reabsorption of tissue fluid

REVISED

If blood vessels were completely closed, no exchange between active cells and the blood would be possible. Ions and small molecules are able to cross capillary walls because the cells forming these walls are very thin (they are squamous cells) and have tiny gaps between them.

The ions and small molecules that leave capillaries form **tissue fluid** that bathes cells in the tissues around capillary networks. Two types of pressure are involved in the formation and reabsorption of tissue fluid: the hydrostatic pressure of the blood and the **oncotic pressure** resulting from the presence of proteins in the plasma.

At the arteriole end of a capillary:
- the hydrostatic pressure is greater than the oncotic pressure, causing a net pressure gradient out of the capillary
- ions and small molecules are pushed down this pressure gradient, out of the blood and into the tissue fluid

> **Tissue fluid** is the fluid that is formed of water, ions and small molecules, which moves from the blood plasma through the walls of capillaries to the surrounding tissues.

> **Oncotic pressure** is the negative water potential of the blood, caused by the presence of proteins in the blood plasma.

- loss of water and small molecules from the blood lowers the hydrostatic pressure within the capillary; as plasma proteins are too large to leave the capillary, the oncotic pressure does not decrease

At the venule end of a capillary:
- the hydrostatic pressure is less than the oncotic pressure, causing a net pressure gradient into the capillary
- water enters the capillary from the tissue fluid down a water potential gradient, i.e. by osmosis
- diffusion gradients result in the diffusion of ions and small molecules from the tissue into the blood plasma

Not all the tissue fluid returns to the capillaries in this way. Some is returned to the blood via veins close to the heart by the **lymph system**.

Now test yourself

21 Give two reasons why the hydrostatic pressure at the venule end of a capillary is less than that at the arteriole end of the same capillary.
22 Why might someone suffering a severe dietary protein deficiency suffer an excess of tissue fluid?

Answers on p. 200

Transport in plants

Multicellular plants are large organisms that have a transport system. Xylem tissue moves water and minerals from the roots to the leaves. Phloem tissue moves assimilates up and down the plant from sources to sinks. Xylem and phloem tissues lie side by side in vascular bundles.

Xylem and phloem tissues

REVISED

Xylem tissue transports water and dissolved inorganic ions. It comprises mainly **xylem vessels**. Mature xylem vessels are well adapted for transporting water by:
- having cell walls that are impregnated with lignin and so can withstand **tension**
- losing their cytoplasm and so there is less impedance to water flow within them
- forming empty, end-to-end cylinders that pass water continuously from root to leaf

Tension is the negative pressure caused by stretching, in this case, a column of water.

Phloem tissue transports the soluble products of photosynthesis, mainly as sucrose with some amino acids. It comprises two types of cell: **sieve tube elements** and **companion cells**. These cells are well adapted for transporting sucrose and amino acids:
- mature sieve tube elements have many pores in their end walls (**sieve plates**), have lost most of their cytoplasm and what little cytoplasm is left runs as strands from cell to cell via pores in the sieve plate
- mature companion cells retain their nucleus and have cytoplasm densely packed with organelles; **plasmodesmata** between the two cells enable each companion cell to control the activities of its adjacent sieve tube element

Plasmodesmata are pores in plant cell walls that enable contact between the cytoplasm of adjacent cells.

The transport of water

There are two main pathways by which water can pass from one plant cell to another:

- The **apoplastic pathway**, in which water moves through spaces between the cellulose molecules within the walls of adjacent cells. The water never enters the cytoplasm of either of the plant cells.
- The **symplastic pathway**, in which water passes by osmosis from the cytoplasm of one cell to the cytoplasm of an adjacent cell via plasmodesmata.

Now test yourself

23 Describe the structures through which a molecule of water must pass if it moves from the permanent vacuole of one plant cell to the permanent vacuole of an adjacent plant cell.

Answer on p. 200

The cohesion–tension model

Figure 4.18 summarises the way in which water is transported from the roots of a plant to its leaves.

> **Exam tip**
>
> The examiner will expect you to use and show understanding of the terms 'cohesion' and 'tension' in an answer involving water transport in plants.

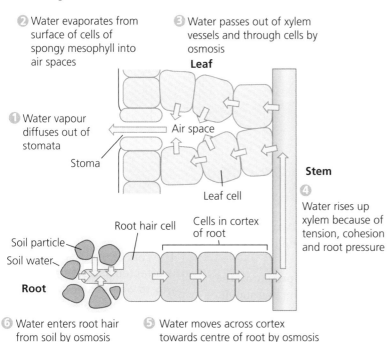

② Water evaporates from surface of cells of spongy mesophyll into air spaces

③ Water passes out of xylem vessels and through cells by osmosis

Leaf

① Water vapour diffuses out of stomata

Air space

Stoma

Leaf cell

Stem

④ Water rises up xylem because of tension, cohesion and root pressure

Root hair cell

Cells in cortex of root

Soil particle

Soil water

Root

⑥ Water enters root hair from soil by osmosis

⑤ Water moves across cortex towards centre of root by osmosis

Figure 4.18 The movement of water through a plant

1 Water vapour in the leaves diffuses down a water vapour gradient from the spaces in the spongy mesophyll through open stomata to the surrounding air. This is called **transpiration**.
2 Water vapour is present in spaces in the spongy mesophyll because water evaporates from the surfaces of these cells.
3 Water lost from cells in the mesophyll is replaced by water in the nearby xylem vessels. As a result of the **cohesion** between water molecules, as one water molecule leaves the xylem it pulls another water molecule behind it.

> **Transpiration** is the loss of water vapour down a water vapour gradient from the leaves through the open stomata to the surrounding air.

4 This pulling stretches the columns of water within each end-on-end tube of xylem vessels. As a result, each column of water is under **tension**. This tension pulls the walls of the xylem vessels inwards slightly.

5 As water is pulled up from the root, it is replaced by water that has crossed the root from the root hair cells.
   ○ Water crosses the root by the apoplast and symplast pathways.
   ○ A layer of cells around the central vascular bundle of the root, called the **endodermis**, has a strip of waxy suberin in each radial cell wall. This is the **Casparian strip** and it prevents water passing any further through the apoplast route. All water must enter the symplast pathway to get past the endodermis.
   ○ The endodermal cells secrete inorganic ions into the xylem vessels, making the water potential in the xylem vessels more negative. Water that has entered the cytoplasm of the endodermal cells then follows into the xylem vessels by osmosis. This increases the hydrostatic pressure within the xylem vessels, causing **root pressure**.

6 Water enters root hair cells by osmosis. The root hair cells are adapted for efficient uptake:
   ○ They have a single extension that enlarges their surface area to volume ratio.
   ○ They use active transport to take up inorganic ions from the soil. This maintains a water potential in the cytoplasm of the root hair cells that is more negative than that of the soil water.

> **Root pressure** is caused by the flow of water by osmosis into the xylem vessels in the root of a plant.

Now test yourself

TESTED ☐

24 Use information from the account of water transport to explain why lignification of the cell walls of xylem vessels adapts them for water transport.

Answer on p. 200

The effect of environmental variables on the rate of transpiration

The variables that affect the rate of transpiration fall into two categories:
● Variables that affect the water potential gradient between the air spaces in the spongy mesophyll and the atmosphere:
   ○ **Air humidity** describes how much water vapour is present in the atmosphere. The water potential gradient between the mesophyll and the atmosphere will be steeper and, therefore, the rate of transpiration will higher when the air is dry than when it is damp.
   ○ **Air temperature** affects the rate of random thermal movement of molecules. The rate of random thermal movement and, therefore, the rate of transpiration will be higher at high temperatures than at lower temperatures.
   ○ **Air currents** move air away from the leaves of a plant. If dry air replaces the moist air around the leaf, the water potential gradient between the mesophyll and the atmosphere will be steeper and, therefore, the rate of transpiration will be higher in windy conditions than in still conditions.

> **Exam tip**
>
> The rate of transpiration is affected in a similar way to washing drying on a line outdoors. Using your understanding of this familiar event might help you in an exam.

- Variables that affect the total area of stomata:
  - ○ **Stomatal opening and closure** affect the area of the exchange surface. The total area of open stomata will be higher in high light intensities than in lower light intensities.
  - ○ **Stomatal density** affects the area of the exchange surface. The greater the density of stomata, the larger the surface area and, therefore, the higher the rate of transpiration.

Now test yourself

TESTED

25 Why will the area of open stomata be greater in high light intensities than in lower light intensities?

Answer on p. 200

The mass-flow hypothesis

REVISED

Translocation occurs from a cell in which a solute is made, called a **source**, to a cell in which it is stored or used, called a **sink**. Movement from source to sink is explained by the mass-flow hypothesis, shown in Figure 4.19.

> **Translocation** is the movement of manufactured solutes in phloem tissue.

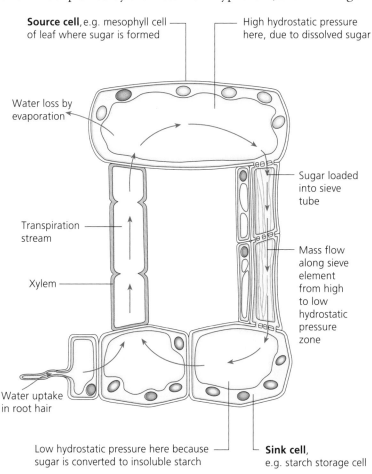

Figure 4.19 The mass-flow hypothesis

At the source

- Sucrose is produced in the mesophyll cells of the leaf during photosynthesis.
- Using active transport, these sugars are loaded into sieve tube elements near the source. This lowers the water potential of these sieve tube

elements (makes their water potential more negative), causing more water to enter the sieve tube elements by osmosis.

- Entry of water into the sieve tube elements increases the volume and, consequently, the hydrostatic pressure of their contents.
- Water that enters the sieve tube elements is replaced by water from xylem vessels near the mesophyll cells.

At the sink

- Sucrose is converted to starch, which is stored. Starch has no effect on the water potential of these cells.
- The removal of sucrose, however, raises the water potential of these cells (makes their water potential less negative), so that water moves out of these cells by osmosis.
- Loss of water from the cells at the sink decreases the volume and, consequently, the hydrostatic pressure of their contents.
- Within sieve tube elements water and its dissolved sucrose moves down a pressure gradient from source to sink, i.e. mass flow occurs.

Evaluating the mass-flow hypothesis

As its name suggests, the mass-flow hypothesis has not yet been fully accepted by scientists. The results of experiments designed to test it have shown the hypothesis has some strengths but also some weaknesses.

Some strengths of the mass-flow hypothesis:
- The contents of sieve tube elements (**phloem sap**) taken from a source have been found to have a higher concentration of sucrose than those taken from a sink. This confirms that the water potential gradient could be formed.
- In infected plants, viruses can be seen to be transported in the phloem from well-illuminated leaves to roots but not from leaves that were in the dark to roots. This suggests that translocation occurs only when sucrose is produced during photosynthesis.
- If a phloem tube element is punctured, such as by inserting a hypodermic needle, phloem sap oozes out, suggesting it is under pressure.

Some weaknesses of the mass-flow hypothesis:
- Organic solutes have been found to move around plants in different directions. If the mass-flow hypothesis is correct, they should all move down a pressure gradient.
- Measurements show that translocation of sucrose and amino acids can occur at different rates within one phloem tube element. If the mass-flow hypothesis is correct, all solutes ought to be translocated at the same rate.

Now test yourself

TESTED

26 Starch grains are often found in cells at a source, such as in mesophyll cells. How does this affect your evaluation of the mass-flow hypothesis?

Answer on p. 200

Exam practice

1 (a) Explain how the structure of its surface membrane affects the exchange of materials between a cell and its environment. [4]

(b) The graph shows the effect of the concentration of an external solution on the rate of entry of two different molecules into a cell. Their rate of entry is the same until point **X** on the graph.

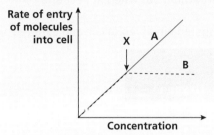

(i) Curve **A** shows the entry of molecules by simple diffusion. Explain the shape of the curve. [1]

(ii) Curve **B** shows the entry of molecules by facilitated diffusion. Explain the evidence from the graph that supports this statement. [2]

2 The diagram represents the blood system of a mammal.

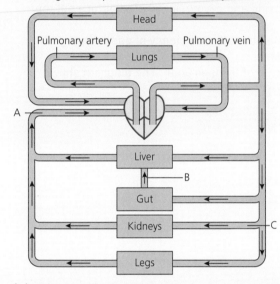

(a) Name each of the blood vessels labelled **A**, **B** and **C**. [3]

(b) Give two ways in which the structure of the pulmonary artery is adapted to its function. [2]

(c) The pulmonary vein contains valves along its length. Describe the role of these valves. [2]

(d) How does the circulatory system shown differ from that of:

(i) an insect [1]

(ii) a fish? [1]

3 (a) Explain how water moves from the central vacuole of one plant cell to the central vacuole of an adjacent plant cell. [2]

(b) Explain how water moves from the root of a tree to its leaves. [4]

(c) Scientists enclosed one leaf on a healthy plant within a clear glass container. They had labelled some of the carbon dioxide in air in this container with radioactive carbon ($^{14}CO_2$). They then left the plant in the light for several hours. After this time, they measured the levels of radioactivity throughout the plant. The table summarises their results.

Position of sample	Measurement of radioactivity/arbitrary units
Bud at tip of stem	3.46
Leaf within glass container	2.34
Stem below glass container	0.48
Root	0.92

Use these results to evaluate the mass-flow hypothesis of translocation in plants. [4]

Exam tip

When making an evaluation of a hypothesis, look for ways in which the data do, and do not, support the hypothesis.

4 (a) Compare and contrast the muscle action that results in breathing out when at rest and breathing out when blowing up a balloon. [3]

(b) The table shows the partial pressures of three gases in inhaled air, alveolar air and exhaled air.

Air	Partial pressure/kPa		
	Carbon dioxide	Nitrogen	Oxygen
Inhaled	0.04	79.8	21.2
Alveolar	5.30	80.3	13.9
Exhaled	3.60	79.6	16.0

(i) Explain why the values for the partial pressure of nitrogen remain almost unchanged. [1]

(ii) Explain why the partial pressures of carbon dioxide and oxygen in exhaled air are not the same as those in the alveolar air. [2]

(iii) What can you conclude from the data in the table about the efficiency of gas exchange in humans? Use calculations to justify your answer. [3]

5 A student investigated the effect of wind speed on the rate of transpiration. The diagram shows the apparatus she used to measure the rate of transpiration.

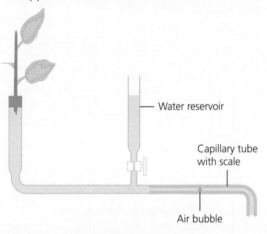

Water reservoir

Capillary tube with scale

Air bubble

The table shows her results.

Wind speed/fan setting	Movement of air bubble/mm
0	6.5
1	13.2
2	16.4
3	18.9

(a) Name the apparatus shown in the diagram. [1]

(b) It is important that there are no air leaks in the apparatus shown. Give one precaution the student should take when setting up this apparatus to prevent air leaks. [1]

(c) Describe the function of the reservoir and tap in the apparatus. [1]

(d) How could the student calculate the rate of transpiration at a fan setting of 0? [3]

(e) Describe and explain the student's results. [4]

→

6 The graph shows the pressures within the left ventricle of someone at rest.

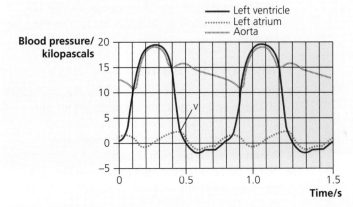

(a) Calculate the heart rate of this person. Show your working. [2]
(b) The atrioventricular valve opened at the point on the graph labelled **V**. Explain the evidence in the graph that supports this statement. [2]
(c) For how long after point **V** did blood flow from the atrium into the ventricle? [1]
(d) The table shows the time taken for a wave of electrical activity to spread from the sinoatrial node to other parts of the heart during one cardiac cycle.

Position in heart	Time/s
Sinoatrial node (SAN)	0.00
Wall of right atrium	0.01
Atrioventricular node (AVN)	0.04
Bundle of His	0.16
Purkyne tissue	0.17

Interpret these data in terms of normal heart function. [4]

Answers and quick quiz 4 online

ONLINE ☐

Summary

Surface area to volume ratio

● As organisms get bigger, their surface area to volume ratio becomes smaller. This affects the efficiency of their exchange surfaces.

Cell transport mechanisms

● The fluid-mosaic model of membrane structure explains the properties of cell surface membranes.
● Small, non-polar molecules can cross cell surface membranes passively by diffusion. Osmosis is a special case of diffusion of water molecules.
● Ions and charged molecules can cross cell surface membranes passively by facilitated diffusion or by active transport.
● Active transport involves the hydrolysis of ATP — the universal energy currency of cells.

Gas exchange

● The gas exchange surfaces of insects, fish, mammals and flowering plants have adaptations that increase the efficiency of gas exchange.
● Fish and mammals increase the efficiency of gas exchange by ventilating their gas exchange surfaces.
● Insects and flowering plants are able to control water loss from their gas exchange surfaces by closure of pores (spiracles or stomata) on their surface.

Circulation

● Circulatory systems enable the mass flow of substances around the body.
● In a closed circulatory system, blood is moved within blood vessels. An open system lacks blood vessels.

- Loss of blood pressure in capillary networks is less in a double circulatory system than in a single circulatory system.
- In addition to mass transport, mammalian blood provides defence mechanisms against infection, including clotting.
- Arteries, veins and capillaries are adapted for their functions.
- A mammalian heart has two pumping systems, providing a circulatory system to the lungs that is separate from that to the body tissues, i.e. a double circulatory system.
- The cardiac cycle involves regular contraction (systole) and relaxation (diastole) of heart muscle.
- Myogenic stimulation of the cardiac cycle is regulated by the sinoatrial node, atrioventricular node, bundle of His and Purkyne tissue. The associated electrical impulses can be displayed using ECG traces.
- Atherosclerosis may lead to the death of cardiac muscle tissue, i.e. myocardial infarction. Genetic and lifestyle factors can increase the risk of atherosclerosis in humans.

Transport of gases in the blood

- Haemoglobin, contained within red blood cells, is an efficient carrier of oxygen from the lungs to the body tissues.
- Fetal haemoglobin and myoglobin have a greater affinity for oxygen than does adult haemoglobin, enabling a fetus and active muscle cells to obtain a rich oxygen supply for aerobic respiration.

Transfer of materials between the circulatory system and cells

- Tissue fluid is forced from capillaries and bathes tissues with water, small soluble molecules and ions. Fluid that is not reabsorbed back into the capillaries returns to the circulatory system via the lymphatic system.

Transport in plants

- Plants have separate systems for the transport of water and organic molecules. Water and dissolved inorganic ions are transported in xylem tissue. Organic molecules are transported in phloem tissue.
- Water can pass from cell to cell via the apoplast pathway and the symplast pathway of plants.
- Cohesion between water molecules ensures that, as water is lost from leaves by transpiration, it is replaced by water that is pulled from xylem tissues in the leaf.
- Transpiration results in tension within xylem vessels that pulls water from the roots to the leaves.
- The rate of transpiration is affected by environmental variables that change the water vapour diffusion gradient from leaves to the atmosphere and by the nature of plant leaves.
- The mass-flow hypothesis provides an explanation of how organic molecules are transported in phloem tissue. The movement of water into phloem cells near photosynthesising tissue creates a hydrostatic pressure gradient that pushes organic molecules to sinks.
- Although there is some experimental evidence that supports it, the mass-flow hypothesis fails to explain the multi-directional flow of organic molecules within a plant.

5 Energy for biological processes

This topic builds on content from Topic 1 (carbohydrates, enzymes and ATP).

- Many metabolic reactions are linked to the hydrolysis of adenosine triphosphate (ATP). The energy released during this hydrolysis drives the link reaction.
- Two processes, respiration and photosynthesis, produce ATP by the enzyme-catalysed condensation reaction:

$$ADP + P_i \xrightarrow{\text{ATP synthase}} ATP$$

Respiration and photosynthesis both involve a complex series of reactions. Many of these:

- are **redox reactions**
- involve **coenzymes**. Three key coenzymes are:
  - **coenzyme A**, which transfers a two-carbon acetate group into the Krebs cycle in respiration.
  - **NAD**, a coenzyme that transfers electrons during cell respiration. When it combines with electrons, it becomes reduced NAD.
  - **NADP**, a coenzyme that transfers electrons during photosynthesis. When it combines with electrons, it becomes reduced NADP.

Exam tip

'P' is the symbol for phosphorus. If you mean 'phosphate' write it in full or use the symbols P_i or PO_4^{3-}.

In a **redox reaction** one molecule is reduced (gains one or more electrons) and another is oxidised (loses one or more electrons).

Coenzymes are organic molecules, other than the substrate, that are needed for enzymes to work. Often they do this by transferring the products of one enzyme-catalysed reaction to another.

Aerobic respiration

Respiration produces ATP, which is the universal energy currency in cells (see page 69). It involves the step-wise breakdown of a **respiratory substrate**. In several of these steps, energy is released that drives the production of ATP from ADP and inorganic phosphate (P_i).

Aerobic respiration uses oxygen but **anaerobic respiration** does not. As a result, aerobic respiration is able to break down a respiratory substrate into CO_2 and H_2O, producing much more ATP per molecule of respiratory substrate than does anaerobic respiration.

Energy transduction from any respiratory substrate to ATP is not 100% efficient; as a result, heat is also released during respiration.

Typical mistake

Students confuse respiration and breathing. Remember, respiration is a process that occurs only within cells.

A **respiratory substrate** is a molecule that is used at the start of respiration. Most commonly, this is a hexose molecule but triglycerides and amino acids can also be used as respiratory substrates.

Exam tip

You will not gain credit if you tell examiners that respiration produces energy. It produces ATP.

Overview of aerobic respiration

REVISED

The stages of aerobic respiration are:

- **glycolysis** — hexose molecules are converted to pyruvate in the cytoplasm
- the **link reaction** — pyruvate is converted to acetate in the mitochondrial matrix
- the **Krebs cycle** — reactions in the mitochondrial matrix produce reduced NAD
- **oxidative phosphorylation** — formation of ATP by molecules embedded in the mitochondrial cristae

Exam tip

Make it clear that aerobic respiration produces more ATP *per molecule of respiratory substrate* than does anaerobic respiration.

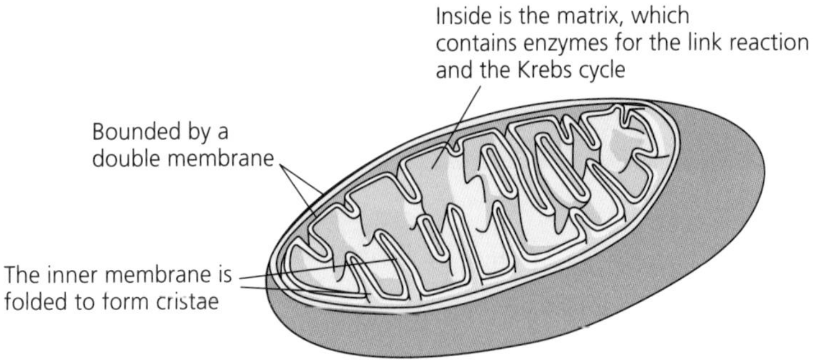

Inside is the matrix, which contains enzymes for the link reaction and the Krebs cycle

Bounded by a double membrane

The inner membrane is folded to form cristae

Figure 5.1 A mitochondrion, in which the link reaction, the Krebs cycle and oxidative phosphorylation all occur

Glycolysis

Hexose molecules cannot enter a mitochondrion. During glycolysis, hexose molecules are converted into pyruvate. This is an anaerobic process and occurs in the **cytosol** of a cell's cytoplasm.

Glycolysis occurs in the three main steps shown in Figure 5.2.

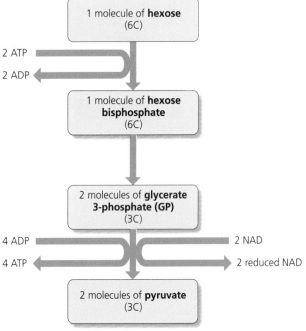

Figure 5.2 The three main steps of glycolysis. The numbers in brackets represent the number of carbon atoms in each molecule

> **Phosphorylation** is the addition of a phosphate group to a molecule, usually making the molecule more reactive.

> **Exam tip**
>
> You are not required to recall most of the reactions involved in respiration.

> **Cytosol** is the semi-fluid part of the cytoplasm that surrounds the organelles.

> **Exam tip**
>
> Don't forget that glycolysis is common to both aerobic respiration and anaerobic respiration.

Now test yourself

TESTED ☐

1. Use your knowledge from Topic 2 to suggest why hexose molecules cannot enter mitochondria.
2. Explain how
   (a) the cristae
   (b) the fluid stroma
   shown in Figure 5.1 adapt a mitochondrion for its function.
3. Suggest the advantage of phosphorylating the hexose molecule in glycolysis.

Answer on p. 200

> **Exam tip**
>
> Keep to the overview: at the end of glycolysis the breakdown of each hexose molecule results in a gain of two molecules of ATP, two molecules of reduced NAD and two molecules of pyruvate.

Link reaction and Krebs cycle

The link reaction

REVISED

Pyruvate molecules pass into the mitochondrial matrix where the reaction shown in Figure 5.3 occurs.

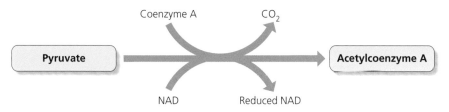

Figure 5.3 **The link reaction**

Now test yourself

TESTED

4 By the end of the link reaction, how many molecules of reduced NAD have been produced from a single molecule of hexose?

Answer on p. 200

The Krebs cycle

REVISED

During the Krebs cycle, a two–carbon acetate is released from each acetylcoenzyme A molecule and is broken down in a cycle of reactions. Each 'turn' of this cycle results in the formation of:

- carbon dioxide
- ATP (by **substrate-level phosphorylation** of ADP)
- reduced NAD

Substrate-level phosphorylation involves the transfer of a phosphate group from one molecule to a molecule of ADP (e.g. $XP + ADP \rightarrow ATP + X$).

Typical mistake

Do not insert an apostrophe into Krebs cycle. Its discoverer was Hans Krebs, not Hans Kreb.

Typical mistake

Many students confuse themselves by trying to memorise the many intermediate compounds in the Krebs cycle. The specification requires you to know only the three bullet points given in the text.

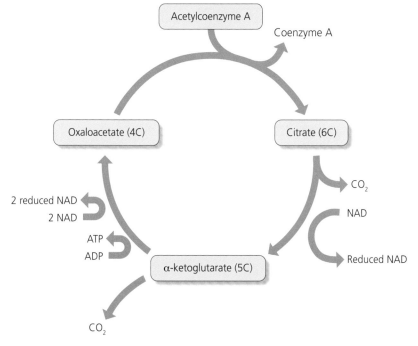

Figure 5.4 **The Krebs cycle**

Oxidative phosphorylation

This final stage of aerobic respiration occurs in the inner mitochondrial membrane and produces ATP by **oxidative phosphorylation**. Proteins embedded in the inner mitochondrial membrane form an **electron transport chain**. These are shown in Figure 5.5.

- At the left of Figure 5.5, a molecule of reduced NAD reacts with a carrier protein in the electron transport chain. The NAD is oxidised as it loses electrons to the protein; protons (H$^+$) are also released.
- These electrons are passed from protein to protein in the electron transport chain.
- Energy released by this electron transport 'powers' the movement of protons though proton pumps into the space between the inner and outer mitochondrial membranes (the **intermembrane space** in Figure 5.5).
- At the end of the electron transport chain, the electrons combine with protons and oxygen to form water. It is this involvement of oxygen that defines this process as 'oxidative'.
- Since oxygen is the last in the transport chain to combine with electrons, it is called the **terminal electron acceptor**.

> **Oxidative phosphorylation** is the electron transport process by which ATP is synthesised from ADP and inorganic phosphate, with oxygen acting as the final electron acceptor.

> **Typical mistake**
>
> Students seem to like abbreviating names and write 'ETC' instead of 'electron transport chain'. Unless a term is given as an abbreviation in the specification or in a question, always write its name in full.

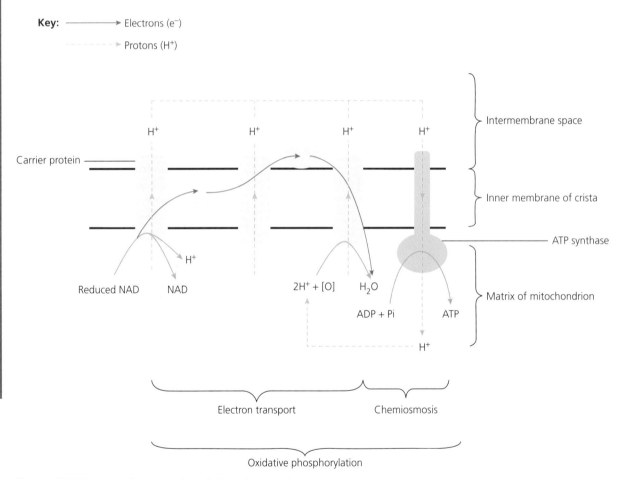

Figure 5.5 **Electron transport and chemiosmosis**

Chemiosmosis and ATP synthesis

REVISED

ATP is synthesised by molecules of ATP synthase embedded in the inner membrane of the cristae.

Pumping of protons into the intermembrane space by the electron transport chain creates a proton diffusion gradient between the intermembrane space and the matrix of the mitochondrion. As a result, protons move from the intermembrane space into the matrix. They do so by facilitated diffusion through molecules of the enzyme ATP synthase.

The diffusion of protons through a molecule of ATP synthase 'powers' the synthesis of ATP from ADP and inorganic phosphate. This is called **chemiosmosis**.

> **Chemiosmosis** is the facilitated diffusion of protons through molecules of ATP synthase and is coupled to the synthesis of ATP.

Now test yourself

TESTED

5 Why can't protons diffuse through the inner mitochondrial membrane other than through the ATP synthase molecules?

Answer on p. 200

> **Exam tip**
>
> Although we refer to electron transport, remember that free protons (H+) are also released during the oxidation of reduced NAD. These protons have a key role during ATP synthesis.

Anaerobic respiration

Anaerobic respiration does not use oxygen. Without oxygen to act as the terminal oxygen acceptor, the electron transport chain cannot function. As a result:

- reduced NAD cannot be re-oxidised to NAD
- without NAD, the link reaction (Figure 5.3) cannot occur

> **Typical mistake**
>
> Students often have trouble describing the role of oxygen in aerobic respiration; it is the final electron acceptor in the electron transport chain.

In animals

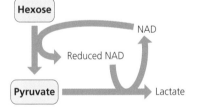

In plants and fungi

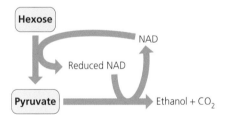

Figure 5.6 **Anaerobic respiration**

As Figure 5.6 shows, glycolysis does continue because reduced NAD is oxidised by pyruvate. This redox reaction produces new end products of respiration:

- **Lactate** in animal cells. It is a build up of this lactate in our muscles that causes muscle fatigue and the pain we associate with cramp.
- **Ethanol** and **carbon dioxide** in the cells of plants and fungi (such as yeast).

Since only glycolysis takes place during anaerobic respiration, only two molecules of ATP are synthesised from the breakdown of one molecule of hexose. This is inefficient compared with the potential 36 molecules of ATP per hexose molecule during aerobic respiration.

> **Typical mistake**
>
> Examiners are often told that anaerobic respiration always produces ethanol. Remember that runners are not drunk at the end of their races!

Now test yourself

TESTED ☐

6 Use Figures 5.2 and 5.6 to justify the statement that anaerobic respiration yields two molecules of ATP per molecule of hexose.

Answer on p. 200

Photosynthetic pigments

Like respiration, photosynthesis involves more than one stage, each comprising a complex series of reactions.

Overall, photosynthesis results in the production of carbohydrates from carbon dioxide and water, with oxygen as a waste product. For this to happen, a photosynthesising organism must first absorb light. Plants, algae and photosynthetic bacteria all possess photosynthetic pigments that absorb light.

Primary and accessory pigments

REVISED ☐

Photosynthetic pigments are lipid-soluble, organic molecules that absorb light. In plant cells, these pigments are located in the **thylakoids** within chloroplasts. Figure 5.7 shows that some of these thylakoids occur in stacks, called **grana** (singular: granum), and some form connections between these stacks.

> **Thylakoids** are the inner membranes of a chloroplast that contain the photosynthetic pigments and enzymes that are essential for the light-dependent stage of photosynthesis.

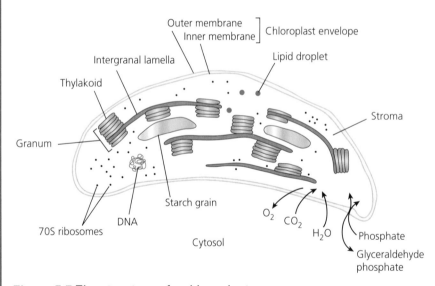

Figure 5.7 **The structure of a chloroplast**

Now test yourself

TESTED ☐

7 What is the link between the photosynthetic pigments being lipid-soluble and being found in the thylakoids?

Answer on p. 201

Most plants have a variety of different plant pigments. The pigment that participates directly in photosynthesis is **chlorophyll a**. For this reason, it is referred to as the **primary pigment**. The other pigments, collectively known as **accessory pigments**, are involved in light capture and include:

- **chlorophyll b**, which is blue green
- **β-carotene**, which is orange
- **xanthophyll**, which is yellow

The accessory pigments are arranged within the thylakoid membranes in funnel-shaped **light-harvesting apparatus**. Here the accessory pigments absorb light and pass it down the funnel to chlorophyll a in a **reaction centre** at the base of the funnel.

> **Accessory pigments** absorb light and use it to activate chlorophyll a.

> The **reaction centre** is the part of a collection of photosynthetic pigments that contains the pigment chlorophyll a.

Absorption and action spectra

REVISED

By shining light of different wavelengths through a chlorophyll suspension and measuring how much is absorbed, you obtain an **absorption spectrum** of that suspension.

The left-hand graph in Figure 5.8 shows the absorption spectra of three components of a chlorophyll suspension. Note that:

- each of the pigments absorbs light across a range of wavelengths
- each pigment has more than one absorption peak
- each pigment has absorption peaks that are different from those of the other pigments
- the chlorophyll mixture has two peaks of light absorption, one at 400–500 nm (violet-blue) and the other at 600–700 nm (orange-red)

By shining light of different wavelengths through a chlorophyll suspension and measuring its rate of photosynthesis, you obtain an **action spectrum** of that plant.

The right-hand graph of Figure 5.8 shows that the rate of photosynthesis is highest in those wavelengths of light that are absorbed collectively by chlorophyll a, chlorophyll b and carotene.

> An **absorption spectrum** represents the extent to which a pigment absorbs light across a range of different wavelengths.

> An **action spectrum** represents the rate of photosynthesis across a range of different wavelengths of light.

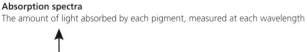

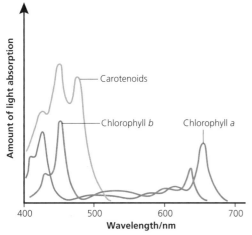
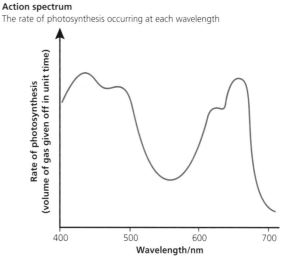

Figure 5.8 The absorption and action spectra of chlorophyll pigments

Now test yourself

TESTED

8 Suggest the advantage to a plant of having more than one photosynthetic pigment.

9 A plant species adapted to living in shady conditions has a combination of chlorophyll pigments that is different from another species adapted to living in bright sunlight. Suggest the advantage of these differences.

Answers on p. 201

Typical mistake

Many students seem to think that each photosynthetic pigment absorbs light at only one wavelength. Figure 5.8 shows that this is clearly not so.

Photosynthesis

Overview of photosynthesis

REVISED

Photosynthesis occurs in two main stages:

● In the **light-dependent stage**, the absorption of light by photosynthetic pigments is linked to the synthesis of **ATP** and the production of **reduced NADP**.

● In the **light-independent stage**, the ATP and reduced NADP from the light-dependent stage are used in a series of reactions that reduce carbon dioxide to produce an organic compound, called **glyceraldehyde phosphate** (**GALP**).

Typical mistake

In their exam answers, students often confuse NAD and NADP. If it helps, remember the **P**: NAD**P** is involved in **p**hotosynthesis.

Exam tip

Students often try to learn so much detail that they seem to lose track of the obvious. It is important that you remember these two bullet points in the overview of photosynthesis.

Light-dependent stage of photosynthesis

REVISED

The light-dependent stage takes place in the thylakoid membranes, where chlorophyll is located. A thylakoid membrane contains a large number of light-harvesting funnels, each with a reaction centre at its base. There are, however, two different types of reaction centre depending on the chlorophylls they contain. They are called photosystem I and photosystem II:

● **Photosystem I** (**PSI**) has a reaction centre that absorbs maximally at 700 nm.

● **Photosystem II** (**PSII**) has a reaction centre that absorbs maximally at 680 nm.

What happens when light strikes chlorophyll in a reaction centre?

When light strikes a chlorophyll molecule, it boosts a pair of electrons out of the chlorophyll molecule. These electrons are passed down an electron transport chain, within the thylakoids.

Non-cyclic photophosphorylation

Non-cyclic **photophosphorylation** produces ATP, reduced NADP and oxygen. It involves the three processes shown in Figure 5.9, which happen simultaneously.

Photophosphorylation is the synthesis of ATP from ADP and inorganic phosphate in the presence of light.

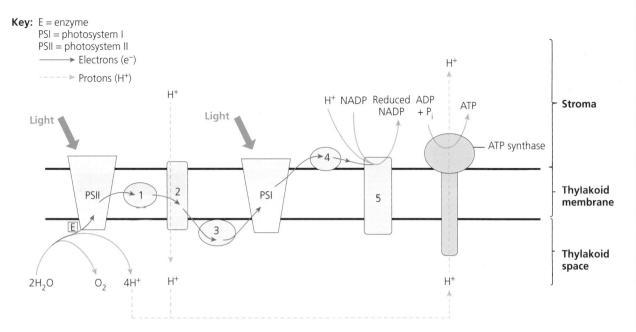

Key: E = enzyme
PSI = photosystem I
PSII = photosystem II
——→ Electrons (e^-)
- - - → Protons (H^+)

Figure 5.9 Non-cyclic photophosphorylation

Process 1 accumulates protons inside the thylakoid – the **thylakoid space**:
- Light strikes PSII, resulting in a pair of electrons leaving a chlorophyll molecule in its reaction centre.
- Each pair of electrons is taken up by a primary electron acceptor (labelled 1 in Figure 5.9) and passed down an electron transport chain (labelled 2 and 3).
- One of the molecules in the chain is a proton pump. Energy released by electron transfer 'powers' the movement of protons from the stroma into the thylakoid space.
- Photosystem II also contains an enzyme (labelled E) that catalyses **photolysis**:

$$2H_2O \rightarrow 4H^+ + 4e^- + O_2$$

- ○ The oxygen is used by the plant during aerobic respiration or diffuses from the plant as a waste product.
- ○ The electrons replace those lost by chlorophyll in photosystem II.
- ○ The protons are used later in photosynthesis.

Process 2 reduces NADP:
- Light strikes chlorophyll in PSI and a pair of electrons leaves a chlorophyll molecule in its reaction centre.
- Each pair of electrons is taken up by a primary electron acceptor and passed down an electron transport chain (labelled 4).
- The electrons, along with protons from the stroma, are used by the embedded enzyme, NADP reductase (labelled 5), to reduce NADP:

$$2H^+ + 2e^- + NADP \xrightarrow{\text{NADP reductase}} \text{reduced NADP}$$

- The chlorophyll in PSI is now 'short' of a pair of electrons. These are replaced by the pair of electrons lost from PSII (hence 'non-cyclic').

> **Photolysis** literally means 'splitting by light'. During the light-dependent stage of photosynthesis water is split, forming protons, electrons and molecular oxygen.

Process 3 synthesises ATP by chemiosmosis:
- Pumping of protons into the thylakoid space creates a proton diffusion gradient between the thylakoid space and the stroma of the chloroplast.
- As a result, protons move from the thylakoid space into the stroma. They do so by facilitated diffusion through molecules of ATP synthase embedded in the thylakoid membrane.
- The diffusion of protons through a molecule of ATP synthase 'powers' the synthesis of ATP from ADP and inorganic phosphate.

Now test yourself

TESTED

10 What is oxygen used for in respiration? Where does it come from in photosynthesis?
11 Which of the numbered structures in Figure 5.9 represents a proton pump?

Answers on p. 201

> **Typical mistake**
>
> Many students tell examiners that plants only respire at night, when they are not photosynthesising. In fact, they respire all the time.

Cyclic photophosphorylation

Cyclic photophosphorylation involves only PSI and results in the production of ATP:
- Light strikes PSI and a pair of electrons leaves a chlorophyll molecule in its reaction centre.
- The pair of electrons is taken up by a primary electron acceptor and is passed down an electron transport chain embedded in the thylakoid membrane.
- A proton pump within the electron transport chain moves protons from the stroma into the thylakoid space.
- The protons diffuse from the thylakoid space back into the stroma through molecules of ATP synthase within the thylakoid membrane, 'powering' the production of ATP from ADP and P_i.
- At the end of the electron transport chain, the electrons return to the molecule of chlorophyll in PSI from which they came (hence 'cyclic').

> **Exam tip**
>
> The processes of chemiosmosis and ATP synthesis occur in respiration and in photosynthesis. This cuts down on the learning you need to do.

Now test yourself

TESTED

12 Name two products of non-cyclic photophosphorylation that are *not* formed during cyclic photophosphorylation.
13 In the early 1950s, Robert Emerson noted that light with a wavelength greater than 690 nm was ineffective for photosynthesis even though it was absorbed by chlorophyll. If he supplemented this with light of shorter wavelength, rapid photosynthesis occurred. What explanation did Emerson suggest for his observation?

Answers on p. 201

Light-independent stage of photosynthesis

REVISED

This stage involves a series of enzyme-catalysed reactions that occur in the stroma of chloroplasts. It can occur in the dark but is dependent on a supply of ATP and reduced NADP produced during the light-dependent reaction.

- Carbon dioxide from the atmosphere is fixed, i.e. it combines in the stroma with a five-carbon compound, called **ribulose bisphosphate** (**RuBP**). This reaction is catalysed by the enzyme ribulose bisphosphate carboxylase (**RUBISCO**).
- The resulting six-carbon compound is unstable and immediately breaks down into two three-carbon molecules of **glycerate 3-phosphate** (**GP**).
- Using ATP and reduced NADP from the light-dependent stage, GP is reduced to form another three-carbon compound, **glyceraldehyde phosphate** (**GALP**).
- Using ATP from the light-dependent stage:
  - some of the GALP is converted into RuBP to continue the Calvin cycle shown in Figure 5.10
  - some of the GALP is converted into other useful compounds, including monosaccharides, triglycerides and amino acids

Exam tip

The abbreviations, RuBP, RUBISCO, GP and GALP are given in the specification, so you may use them in exam answers.

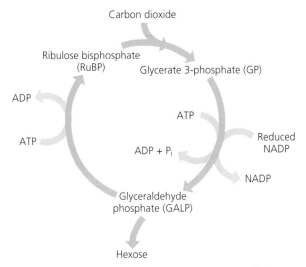

Figure 5.10 The light-independent stage of photosynthesis (also known as the Calvin cycle)

Now test yourself

TESTED

14 Although the light-independent stage does not require light directly, it will not continue for very long in the absence of light. Explain why.

Answer on p. 201

Factors that limit the rate of photosynthesis

REVISED

A number of environmental factors affect the rate of photosynthesis. The three main factors are:
- **carbon dioxide concentration**, because carbon dioxide is a substrate in the light-independent stage
- **light intensity**, because the absorption of light is essential for the light-dependent stage
- **temperature**, because many of the reactions of photosynthesis are catalysed by enzymes

All of these factors are permanent features of a plant's environment and so, collectively, affect the rate of photosynthesis. The one that is in 'least supply' will have the greatest influence on the overall rate. It is the **limiting factor**.

A **limiting factor** is the environmental factor that is in least supply for any process that is affected by several environmental factors.

Figure 5.11 shows the effect of carbon dioxide concentration, light intensity and temperature on the rate of photosynthesis.

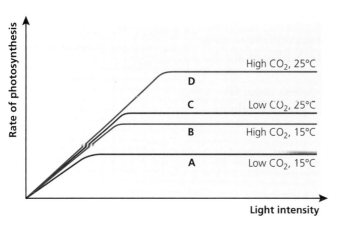

Figure 5.11 The effect of light intensity on the rate of photosynthesis at two different carbon dioxide concentrations and two different temperatures

Exam tip

Students often wonder why they fail to gain marks when they refer to 'light' or 'CO$_2$' being limiting factors. You must use appropriate terminology by referring to 'light *intensity*' or 'CO$_2$ *concentration*'.

Now test yourself

TESTED ☐

15 How can you tell that light is the limiting factor in the early part of curves A to D in Figure 5.11?

Answer on p. 201

In the region of the curves where the rate of photosynthesis remains constant, increasing light intensity does not result in an increase in the rate. In these regions:
- the limiting factors in curve **A** are both carbon dioxide concentration and temperature, because increasing either results in an increase in the rate
- the limiting factor in curve **B** is temperature, because increasing the temperature to 25°C (curve **D**) increases the rate
- the limiting factor in curve **C** is carbon dioxide concentration, because increasing it (curve **D**) results in an increase in the rate

Typical mistake

Seeing the plateaus in curves **A** to **D** in Figure 5.11, many students report that photosynthesis has stopped. The plateau indicates that the rate of photosynthesis has become constant, not that photosynthesis has stopped.

Now test yourself

TESTED ☐

16 Market gardeners often use paraffin heaters in their glasshouses. Give *two* ways in which this will increase the rate of photosynthesis of their crop plants.
17 Suggest *three* factors within a plant that might limit its rate of photosynthesis.

Answers on p. 201

Exam practice

1 Oxygen is used in aerobic respiration and produced in photosynthesis.
   (a) What is the role of oxygen during aerobic respiration? [2]
   (b) What is the source of oxygen released during photosynthesis? [2]
2 (a) Give *two* different ways in which the hydrolysis of ATP benefits cell metabolism. [2]
   (b) Identify precisely the location(s) of ATP synthesis in a mesophyll cell from a plant leaf. [2]
   (c) Describe the chemiosmosis model of ATP synthesis. [4]
3 (a) Precisely where do the following stages of photosynthesis occur in a plant cell?
      (i) the light-dependent stage
      (ii) the light-independent stage [2]
   (b) A scientist investigated the process of photosynthesis using cultures of single-celled algae:
      ● She provided a liquid suspension of algae with carbon dioxide containing radioactively labelled carbon ($^{14}CO_2$) and allowed the algae to photosynthesise.
      ● After some time, she removed the light source and left the algae in the dark.
      ● Throughout the experiment, she took samples from the suspension of algae and found which substances in the algae were radioactively labelled.
      The diagram summarises her results.

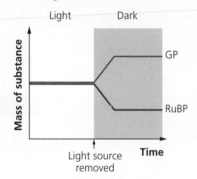

      Explain the scientist's results. [6]
4 The diagram shows the apparatus used by a group of students to investigate respiration using cultures of yeast — a single-celled fungus.

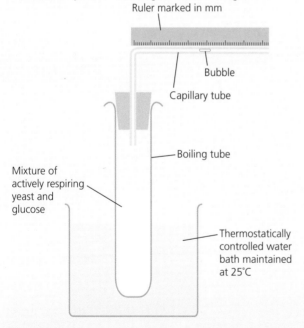

   The capillary tube contained a liquid within which was an air bubble.
   ● The students positioned the air bubble so that it was at the middle of their ruler, i.e., at the 150 mm mark.
   ● They used the ruler to measure the position of the bubble of air every minute for 10 minutes.
   ● A distance less than 150 mm indicated a movement of the air bubble towards the tube; a distance greater than 150 mm indicated movement away from the tube.

The students repeated their experiment 10 times. Their results are shown in the table.

Time/minutes	Mean distance of air bubble from tube/mm
0	150
1	150
2	150
3	+ 152
4	+ 155
5	+ 159
6	+ 163
7	+ 166
8	+ 171
9	+ 125
10	+ 128

(a) Explain why the students placed the tube in a water bath. [2]

(b) Explain why the students repeated their experiment 10 times. [2]

(c) Suggest how the students could prevent air leaks in their apparatus. [1]

(d) Explain the students' results. [5]

Answers and quick quiz 5 online

ONLINE

Summary

ATP

- Many metabolic reactions are linked to the hydrolysis of adenosine triphosphate (ATP). The energy released during this hydrolysis is used to drive the link reaction.
- ATP is synthesised from ADP and phosphate (P_i) during respiration and during the light-dependent stage of photosynthesis.

Respiration

- The reactions of glycolysis occur in the cytoplasm of cells. During these reactions, hexose molecules are hydrolysed to form pyruvate, with the production of some ATP and reduced NAD. Glycolysis is common to aerobic and anaerobic respiration.
- The link reaction and the Krebs cycle take place in the matrix of mitochondria. The link reaction results in the formation of acetylcoenzyme A from pyruvate. Acetylcoenzyme A enters the Krebs cycle, which results in the formation of ATP, reduced NAD and carbon dioxide.
- The electron transport chain takes place in the inner membranes of mitochondrial cristae. During this process:
  - electrons are transported along a chain of carriers, the final one of which is oxygen
  - the transport of electrons drives proton pumps that pump protons into the space between the two mitochondrial membranes
  - facilitated diffusion of these electrons back into the mitochondrial matrix through molecules of ATP synthase drives the formation of ATP from ADP and P_i — a process called chemiosmosis

Photosynthesis in plants

- The chloroplasts of plants have a range of photosynthetic pigments within their lamellae. The range of light wavelengths that each pigment absorbs is its absorption spectrum. The range of wavelengths across which photosynthesis occurs is the chloroplast's action spectrum.
- Photosynthesis comprises two stages: the light-dependent stage and the light-independent stage.
- The light-dependent stage of photosynthesis results in the formation of ATP and reduced NADP, both of which are used in the light-independent stage. Light causes:
  - photolysis — the splitting of water to produce protons, electrons and molecular oxygen
  - excited electrons to leave molecules of chlorophyll a
- The transport of electrons along an electron transport chain within the lamellae results in the formation of reduced NADP and drives proton pumps that pump protons into the space inside the thylakoids.
- Facilitated diffusion of these electrons back into the stroma of the chloroplast through molecules of ATP synthase drives the formation of ATP from ADP and P_i — a process called chemiosmosis.
- The light-independent stage of photosynthesis occurs in the stroma of chloroplasts. In this process:
  - carbon dioxide combines with RuBP to form two molecules of GP
  - using ATP and reduced NADP from the light-dependent stage, GP is reduced to GALP. Some GALP is used to synthesise monomers, such as glucose and amino acids; some is used to regenerate RuBP.
- Carbon dioxide concentration, light intensity and temperature can all act as limiting factors on photosynthesis.

6 Microbiology and pathogens

This topic builds on content from Topic 2 (eukaryotes, prokaryotes and viruses) and Topic 4 (circulation).

Microbial techniques

Culturing microorganisms

REVISED

When handling cultures of microorganisms, you must use **aseptic techniques** to prevent:

- potentially harmful microorganisms escaping from your cultures into the air
- microorganisms in the air entering and contaminating your cultures

Aseptic techniques

The most common of these techniques are summarised in Table 6.1, along with the reasons for their use.

> **Aseptic techniques** are safety precautions, used when handling microorganisms, which reduce the risk of contaminating pure cultures and of infecting humans.

Table 6.1 **Examples of aseptic techniques**

Aseptic technique	Purpose of technique
1 Thoroughly wash hands before and after the laboratory procedure	Washing before reduces the risk of contaminating cultures with (potentially harmful) skin bacteria Washing after reduces the risk of leaving the laboratory with bacteria from the cultures remaining on the skin
2 Avoid all hand-to-mouth contact	Prevents microorganisms being transferred into the body via the mouth
3 Clean the surface of bench to be used using antiseptic solution	Reduces risk of contamination of pure cultures by microorganisms on the bench
4 Have a container of antiseptic solution on area of bench to be used	Antiseptic immediately available to: ● deal with any accidental spillages ● dispose of glass pipettes that have been contaminated when transferring liquid cultures
5 Have a lighted Bunsen burner easily accessible on the bench	Used to flame-sterilise dry glassware and inoculating loops Creates an upward current of warm air that removes microorganisms from the bench area
6 Flame-sterilise inoculating loop immediately before and after use	Kills any microorganisms on the loop
7 Flame-sterilise top of glass culture bottles immediately before and after use	Reduces risk of contaminating culture with airborne microorganisms and release of microorganisms from culture
8 Keep hold of top/cap of glass culture bottles when flame-sterilising and when transferring microorganisms	Avoids putting top/cap onto bench, thus reducing risk of transferring microorganisms between container and bench

Aseptic technique	Purpose of technique
9 When using an inoculating loop with an agar culture, lift the lid of the Petri dish at an angle and only sufficiently to allow manipulation of the inoculating loop	Reduces risk of microorganisms escaping from the agar into the air or of other microorganisms from the air contaminating the agar
10 After inoculating an agar plate, use sticky tape to seal the lid to the base of the Petri dish	Reduces the risk of microorganisms escaping if the lid became dislodged
11 At the end of a session in a college or school laboratory, incubate cultures at temperatures well below 37°C	Avoids growth of pathogenic microorganisms that grow best at body temperature
12 At the end of a laboratory session, place all equipment and unwanted cultures in an autoclave at 121°C for at least 15 minutes	Kills all organisms on the apparatus and in the unwanted cultures Disposal of the unwanted cultures is now safe

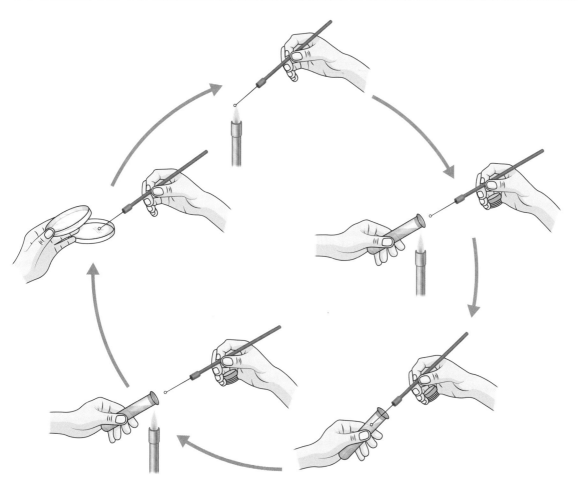

Figure 6.1 Aseptic techniques being used during the transfer of a small sample of microorganisms from a liquid culture to an agar plate

Now test yourself

TESTED ☐

1 Use the numbers in Table 6.1 to list the aseptic techniques being demonstrated in Figure 6.1.
2 A lighted Bunsen can be used to flame-sterilise dry glassware. Suggest why it would *not* be safe to flame-sterilise a glass pipette that had been used to transfer a liquid culture of microorganisms.

Answers on p. 201

Exam tip

In addition to naming them, ensure that you can explain the reasons for a number of aseptic techniques.

Different types of culture media

Microbiologists usually start a culture by:

- swabbing an area with, for example, a sterile cotton bud and exposing the surface of the bud to a sterile culture medium, or
- transferring an **inoculum** from an existing culture to a fresh culture medium

Many different culture media exist. These can be:

- **liquid**, growing cultures referred to as 'broth' cultures
- **solid**, growing cultures that form discrete colonies, which can be removed using an inoculating loop
- **non-selective**, allowing the growth of a broad range of microorganisms
- **selective**, allowing the growth of a narrow range (or even a single species) of microorganisms

Depending on the purpose of the microorganism, microbiologists can grow:

- **batch cultures**, which have a fixed volume of growth medium (e.g. agar in a Petri dish)
- **continuous cultures**, in which spent liquid medium is continuously removed from the culture vessel and replaced with fresh liquid culture medium

> **Inoculum** refers to the small sample taken from one culture and transferred, using aseptic techniques, to a new sterile medium.

Now test yourself   TESTED ☐

3 Which type of culture, batch or continuous, would a microbiologist choose when:
   (a) culturing microorganisms that produce a useful product?
   (b) storing a sample of bacteria in a refrigerator?

Answers on p. 201

Isolating an individual species from a mixed bacterial culture using streak plating

If you were to swab your skin or an area of the laboratory and use the swab to **inoculate** an agar plate, you would end up with a mixture of several species of microorganism. You can isolate one type of microorganism from this using streak plating. By repeatedly spreading the inoculum in the directions shown in Figure 6.2, you are diluting the inoculum to such an extent that colonies are formed from individual cells. You can then obtain a pure culture by removing a sample from a single colony and inoculating it into a fresh, sterile medium.

> You **inoculate** a sterile medium when, using aseptic techniques, you transfer an inoculum into it.

Figure 6.2 **The streak-plating technique**

Now test yourself   TESTED ☐

4 Add an arrow to Figure 6.2 to show the point at which you would expect to obtain individual colonies.
5 You would use aseptic techniques during the streak-plating technique. Give one aseptic technique you would use involving the inoculating loop and one aseptic technique you would use involving the Petri dish.

Answers on p. 201

Measuring the growth of bacterial cultures

REVISED

To measure the growth of a bacterial culture, you need to find the number of cells in a known volume of liquid culture at different time periods after inoculation.

Making a serial dilution of the bacterial culture

In any bacterial culture, the number of cells is far too large to count accurately. Before making an estimate of the number, you must:
- make a serial dilution of the culture
- find one dilution that enables you to estimate the number of bacteria
- calculate the mean number of cells in several samples of that dilution

Knowing the volume of the samples and the dilution factor, you can then calculate the number of bacteria in the original undiluted culture.

Figure 6.3 shows how, using aseptic techniques, you would make a serial dilution of a culture.

> **Exam tip**
>
> Be sure you can explain that an appropriate dilution is a compromise between having a low enough number of microorganisms to count and a high enough number to provide a reliable sample of the culture.

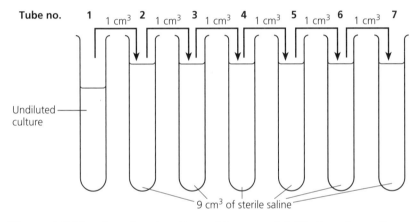

Figure 6.3 The steps involved in making a serial dilution of a culture of microorganisms

Now test yourself

TESTED

6 If the density of bacterial cells in tube **1** of Figure 6.3 is 1.0, what is the density of bacterial cells in tube **4**? Give your answer in standard form.

Answer on p. 201

You can make two types of cell count:
- A **total count** includes all the cells in the culture, whether alive or dead.
- A **viable count** includes only the living cells in the culture.

Making total counts

Three methods are available for making total counts.

Direct count using a haemocytometer

Figure 6.4 shows the grid etched onto a **haemocytometer**. Microbial cells are shown on this grid. Since they can easily be counted, these cells are from a suitable dilution of the original culture. By convention, you should *not* include in your count any cell that is touching or overlapping the middle of the three lines at the bottom and left-hand side of a $0.2\,\text{mm} \times 0.2\,\text{mm}$ square.

> A **haemocytometer** is a glass slide on which is etched a grid of known dimensions that, when covered by a coverslip, encloses a known volume of liquid.

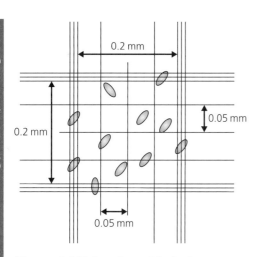

Figure 6.4 Using the grid of a haemocytometer to make a direct count of cells

Exam tip

The following mnemonic might help you to remember to exclude cells touching the left and bottom lines of the 0.2 mm × 0.2 mm haemocytometer square: **LBW** (**l**eft and **b**ottom **w**ithheld).

Now test yourself

TESTED

7 How many cells should you count in the 0.2 mm × 0.2 mm square in Figure 6.4?

8 If the gap between the coverslip and the haemocytometer slide is 0.1 mm, calculate the volume of culture covering the 0.2 mm × 0.2 mm square in Figure 6.3. Give your answer in standard form.

Answers on p. 201

Indirect count measuring dry mass of cells

Using aseptic techniques, you filter a sample of known dilution of the culture through a pore size of $0.2\,\mu m$. You then remove water from the sample by heating the filter paper to constant mass. By subtracting the final mass from the original mass of the filter paper, you can calculate the dry mass of microorganisms in a known volume of culture.

Indirect count measuring turbidity of diluted culture

A liquid sample containing many microorganisms is cloudier than one containing fewer microorganisms. The 'cloudiness' can be measured using a simple **colorimeter**. By placing samples of culture into a colorimeter, you can measure either the absorbance or transmission of light by the cells in the culture. One drawback of this technique is that you have to use a direct method, for example a haemocytometer, to produce a calibration curve that relates absorbance or transmission to the actual number of cells in the sample.

A **colorimeter** measures the absorbance of light by, or transmission of light though, a liquid.

Making viable counts using spread plating

Having prepared a serial dilution of the original culture, you use aseptic techniques to:
- inoculate agar plates with a sample of known volume from each dilution
- spread each inoculum across its agar plate
- incubate the agar plates until a colony has grown from each cell in the inoculum
- choose plates that have a small enough number of colonies to count but a large enough number to provide a valid sample size
- use the number of colonies per plate to calculate the number of viable cells in the original culture

Now test yourself

TESTED ☐

9 Other than the number of colonies per plate, what else would you need to know to calculate the number of cells in the original culture?

Answer on p. 201

The bacterial growth curve

REVISED ☐

By measuring the viable count of a bacterial culture over time, you can produce a population growth curve. Figure 6.5 shows a typical growth curve for a bacterial population grown in batch culture. The growth curve has four phases:

- The **lag phase**, during which the bacteria do not reproduce but are active, producing new ribosomes and transcribing genes that will enable them to exploit the new medium.
- The **log phase**, during which the bacteria divide by binary fission at their maximum rate. Population growth is now exponential (or logarithmic, hence the term 'log' phase).
- The **stationary phase**, during which cells die at the same rate as new cells are formed by binary fission. Cell death results from depletion of nutrients in the medium or the accumulation of toxic waste products of bacterial metabolism.
- The **death phase**, during which more and more cells die.

> **Typical mistake**
>
> Students often tell examiners that bacteria divide by mitosis. Remember that mitosis only occurs in eukaryotic cells.

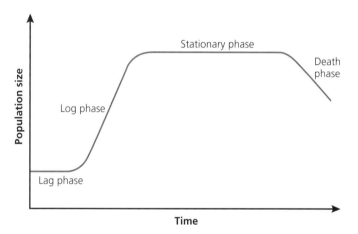

Figure 6.5 A typical growth curve for bacteria grown in batch culture

Use of a growth curve to calculate the exponential growth rate constant

The exponential growth rate constant (k) is the rate at which bacteria grow during the log phase of the growth curve. You can calculate the growth rate constant using the following formula:

$$k = \frac{\log_{10} N_t - \log_{10} N_0}{0.301 \times t}$$

where N_0 is the number of cells at the start, N_t is the number of cells at the later time and t is the time interval.

Bacteria as pathogens

Many bacteria are agents of infection, i.e. are **pathogens.** They harm their hosts by:

- invading and destroying host tissues — for example, *Mycobacterium tuberculosis*, which causes tuberculosis in humans, destroys lung tissue
- producing **toxins**

Endotoxins are released by bacteria only on the death and lysis of the bacterial cells that contain them. The endotoxins of *Salmonella enterica* cause local inflammation in the gut, whereas the endotoxins of *Salmonella typhi* cause the widespread symptoms of typhoid fever.

Exotoxins are secreted by living bacteria and poison the host's tissues. For example, species of *Staphylococcus*, including the strain of *Staphylococcus aureus* that has become resistant to methicillin (i.e, **m**ethicillin-**r**esistant *Staphylococcus aureus*, MRSA), secrete exotoxins.

Action of antibiotics

Antibiotics can affect bacteria in one of two ways:

- **Bacteriostatic antibiotics** prevent the multiplication of bacteria but do not kill them. For example, penicillin is a bacteriostatic antibiotic. It inhibits an enzyme that is essential for the production of the cell wall in Gram-positive bacteria (page 30).
- **Bactericidal antibiotics** kill bacteria. For example, tetracycline is a bactericidal antibiotic. It prevents the formation of ribosomes in both Gram-positive and Gram-negative bacteria.

Antibiotic resistance

The structure of their cell walls (pages 30–31) makes Gram-negative bacteria resistant to penicillin. This is called **primary resistance**. In recent years, many bacteria that were once susceptible to an antibiotic have developed resistance to it (pages 60–61). This is called **secondary resistance**.

Now test yourself

TESTED ☐

10 Give the difference(s) between the structure of the cell wall of Gram-positive bacteria and that of Gram-negative bacteria.

Answer on p. 201

Development of secondary resistance to antibiotics

REVISED ☐

Secondary resistance is an example of natural selection. It follows the following sequence of events:

- A chance mutation occurs in a single bacterial cell.
- As a result of the mutation, a new form of the gene (an **allele**) enables the bacterium to resist the effects of an antibiotic. This could involve:
  - ○ a change in the uptake of the antibiotic by the bacterium
  - ○ production of an enzyme that modifies or inactivates the antibiotic
  - ○ development of a new metabolic pathway that bypasses the reaction affected by the antibiotic

---

Exam tip

Examiners could combine a question about pathogens with content from Topic 2, so you need to make these links during your revision.

Pathogens are agents of disease. They include microorganisms but also viruses and proteins called prions.

A **toxin** is a poison produced by one organism that has a harmful effect on another organism.

Antibiotics are substances that, in low concentrations, kill or inhibit the growth of microorganisms.

Typical mistake

Many students give a generic response to questions about natural selection, failing to relate it to the organisms in the question. As a result, they fail to gain marks.

Exam practice answers and quick quizzes at **www.hoddereducation.co.uk/myrevisionnotes**

- In the presence of the relevant antibiotic, susceptible bacteria will either be killed or fail to reproduce. Only the resistant bacterium will reproduce, passing the new allele on to its offspring.
- As a result, the frequency of the allele conferring resistance will become common.

Controlling the spread of antibiotic resistance

REVISED

People infected by antibiotic-resistant bacteria are often taken into hospital for treatment. Here they could infect other patients with the antibiotic-resistant bacterium. The risk of this cross-infection can be reduced by:
- isolating the infected person
- hospital staff and visitors washing their hands and using a skin disinfectant on leaving an infected patient
- hospital staff wearing gloves and aprons that they dispose of immediately after handling infected patients

Now test yourself

TESTED

11 Staff in many hospitals are discouraged from wearing neck ties or scarves. Suggest why.

Answer on p. 201

Although the use of antibiotics does not cause mutations that lead to antibiotic resistance in bacteria, the risk of bacteria evolving antibiotic resistance can be reduced by:
- restricting the use of antibiotics by:
  - persuading doctors not to prescribe antibiotics unless absolutely necessary
  - restricting over-the-counter availability of antibiotics, which is common in many countries
  - restricting the use of antibiotics in animal feeds
- ensuring that patients complete their course of antibiotics, so that all bacteria are killed before a chance mutation occurs
- developing new antibiotics — which is an extremely expensive process

All the above measures involve either persuading people to change their behaviour, increased expenditure or both. As a result, they have proved hard to implement.

Typical mistake

It is wrong to write that antibiotics *cause* mutations conferring resistance. Mutations occur by chance.

Now test yourself

TESTED

12 In the UK, the NHS has a target for the primary care sector to reduce the issue of prescriptions for antibiotics by 4%. Explain why.

Answer on p. 201

Other pathogenic agents

Bacteria are not the only pathogens that affect humans, their livestock or their crops. Table 6.2 summarises the effects of three other pathogens that you need to know.

Table 6.2 Stem rust fungus, the influenza virus and the malarial parasite are pathogens

Pathogen	Transmission	Mode of infection	Pathogenic effect
Stem rust fungus (*Puccinia graminis*)	Airborne spores of fungus	Spores germinate and produce a mass of thread-like **hyphae** that penetrate the stem tissue of the host wheat plant	Secretes enzymes that digest host tissues, causing stem to fall over, resulting in a smaller wheat crop and increased risk of infection by other pathogens
Influenza ('flu') virus	Droplet infection (coughs and sneezes) from one person to another	Taken into cells lining bronchi and bronchioles by endocytosis	Causes lysis of infected cells and release of toxins that have widespread effects in the body
Malarial parasite (*Plasmodium* spp.)	Parasite enters body in saliva of a female *Anopheles* mosquito as it takes a blood meal	Enters liver cells where rapid reproduction occurs	Parasite released by lysis of liver cells enter red blood cells, reproduce and cause lysis of red blood cells releasing further parasites and toxins into blood

Now test yourself

TESTED ☐

13 Explain why infection by stem rust fungus would increase the risk of a plant becoming infected by other pathogens.

Answer on p. 201

Problems of controlling endemic diseases

Malaria is an **endemic** disease in many countries. The World Health Organization (WHO) estimates that about 400 million people are infected with malaria, of whom 1.5 million die each year. Although malaria is a health priority, its control raises many social, economic and ethical implications. Some of these are summarised in Table 6.3.

> **Endemic** diseases occur permanently in particular populations.

Now test yourself

TESTED ☐

14 Express 400 million in standard form.

Answer on p. 201

Table 6.3 Issues raised by control methods for endemic malaria

Method that might reduce the incidence of malaria	Issues raised
Drain wetlands where mosquitoes breed	1 Not possible to drain lakes from which many people derive their food and their income 2 Large lakes span more than one country, necessitating inter-governmental cooperation
Spray areas with insecticide to kill *Anopheles* mosquito	3 Insecticides could harm populations of beneficial insects or accumulate in food chains 4 Vast areas of land are involved, requiring inter-governmental cooperation or cooperation between warring groups
Cover beds with nets sprayed with insecticide	5 Some mosquito populations have evolved the habit of biting during the day rather than at night
Release sterilised males that mate with females that then fail to lay fertile eggs	6 Releasing more insects in order to reduce the insect population is counterintuitive 7 Large-scale education programmes are needed to convince local populations of the logic of this method
Vaccination	8 The development of vaccines is expensive 9 *Plasmodium* shows great antigenic variability 10 Malaria is endemic in developing countries that lack the infrastructure to deliver effective vaccination programmes

Now test yourself

TESTED

15 *Plasmodium* shows great antigenic variability. Explain what this means and why it causes a problem in developing vaccines.

Answer on p. 201

Revision activity

Arrange the numbers of each issue in Table 6.3 into three groups: social, economic and ethical.

Response to infection

Cells that are foreign to your own body carry foreign **antigens**. Your immune response depends on the ability of a number of **leucocytes** (Figure 6.6) to 'recognise' these foreign antigens.

Antigens are macromolecules (such as proteins, polysaccharides, lipoproteins or glycolipids) that cause an immune response in the body.

Leucocytes are white blood cells, all of which are involved in a mammal's response to infection.

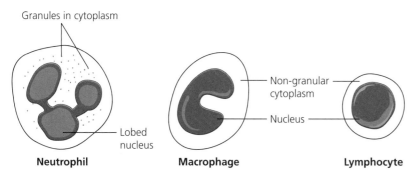

Figure 6.6 Three leucocytes: a neutrophil, a macrophage and a lymphocyte

Phagocytosis

The first leucocytes to be involved in your immune response are **phagocytic cells**, which ingest and destroy foreign microorganisms. Two phagocytic leucocytes — neutrophils and macrophages — are shown in Figure 6.6. They both have surface receptors for antibodies and so engulf bacterial cells that are coated by your own antibodies.

- **Neutrophils** make up about 60% of your blood-borne leucocytes. Once it has ingested a bacterial cell, a neutrophil kills and digests it.
- **Macrophages** are cells that are found in your body tissues, especially in your lungs, liver and lymph nodes. In addition to ingesting foreign cells, they ingest any of your own cells that are dead or dying. Once they have ingested a bacterial cell, they:
  - hydrolyse the proteins of the ingested cell into peptide fragments, which they then display on their own cell surface membranes. In this condition, they are called **antigen-presenting cells**.
  - release chemical messengers, called **cytokines**, that stimulate other immune responses

> **Antigen-presenting cells** engulf pathogens by phagocytosis and display protein fragments from the pathogen on their own cell surface membranes.
>
> **Cytokines** are small proteins that are involved in cell signalling.

Lymphocytes and clonal selection

Lymphocytes are small, nucleated cells (Figure 6.6) that give you immunity against specific microorganisms. Over your lifetime, you produce millions of lymphocytes with different surface receptor proteins.

Figure 6.7 shows what happens when, by chance, the receptor protein of one lymphocyte binds to a complementary antigen that is displayed by an antigen-presenting cell.

- The lymphocyte becomes sensitised and divides repeatedly to produce a clone of cells.
- Most of the cells become activated lymphocytes; a few remain as inactivated memory cells.

This process is called **clonal selection** because:
- all the cells produced by the sensitised lymphocyte are genetically identical (are a clone)
- only one lymphocyte, from the millions you produce with different receptors, is 'selected'

> **Clonal selection** involves the repeated mitotic divisions of a sensitised lymphocyte to produce a large number of genetically identical cells, most of which are active lymphocytes but a few of which are inactive memory cells.

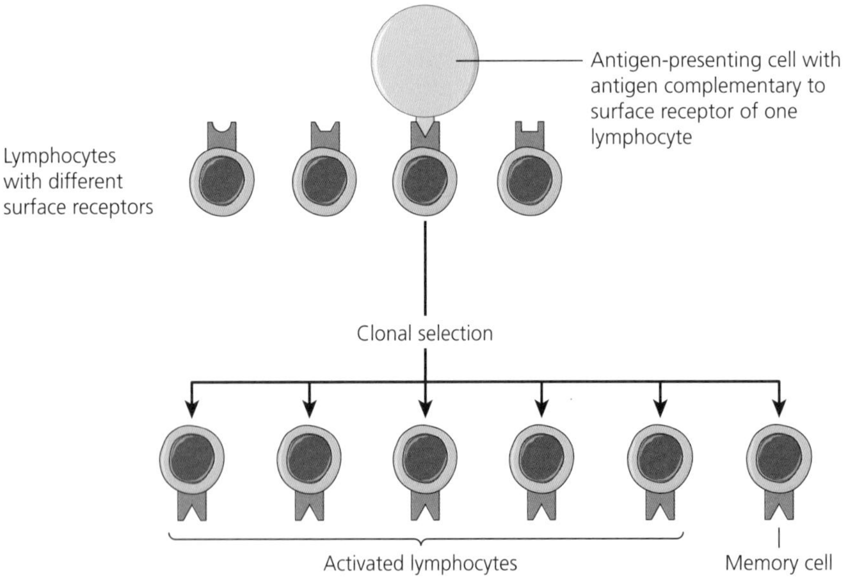

Figure 6.7 Clonal selection produces activated lymphocytes and inactive memory cells

Now test yourself

TESTED ☐

16 A single pathogen is likely to cause the clonal selection of several lymphocytes. Suggest why.

Answer on p. 201

There are two main types of lymphocyte. Both are formed in bone marrow but differentiate in different tissues:

- B lymphocytes (**B cells**) differentiate within the **b**one tissue in which they were produced. They provide the **humoral immune response** against antigens and pathogens in body fluids.
- T lymphocytes (**T cells**) migrate from the bone tissue in which they were produced and differentiate in the **t**hymus gland. They are responsible for the **cell-mediated immune response** against abnormal cells and pathogens inside living cells.

The cell-mediated immune response

REVISED ☐

The cell-mediated immune response involves the interaction of macrophages and T cells, as follows:

- **Antigen presentation** occurs when a macrophage engulfs a foreign cell, breaks it into peptides fragments and displays these peptides as antigens on its own cell surface membrane.
- **T cell activation** follows the binding of a T cell with a surface receptor that is complementary to the antigen on an antigen-presenting cell.
- **Clonal selection** of the T cell then follows, producing active T cells and memory cells.
- T cells with one type of surface receptor (called CD4 receptors) produce **T helper cells** that release cytokines that:
  - ○ stimulate activation and clonal selection of B cells
  - ○ regulate the activity of T killer cells
- T cells with a second type of surface receptor (called CD8 receptors) produce **T killer cells** that destroy any cell infected by the pathogen that has the antigens displayed by the antigen-presenting cell. They do this in one of several ways:
  - ○ rupturing the cell surface membrane of the infected cell
  - ○ releasing lymphotoxins that poison the infected cell
  - ○ activating genes within the infected cell that cause programmed cell death (apoptosis)
- A few T cells in each clone do not become active but remain as inactive **memory cells**.

The humoral immune response

REVISED ☐

The humoral immune response involves the release of **antibodies**. It involves the following interaction of antigen-presenting cells, T helper cells and B cells:

- **B cell sensitisation** occurs following binding of the antigen on an antigen-presenting cell with a surface receptor on a B cell that is complementary to that antigen.
- **B cell activation** follows when an activated T helper cell binds to the sensitised B cell and begins to secrete **cytokines** that stimulate clonal selection of the activated B cell.

> **Antibodies** are a type of glycoprotein (called immunoglobulins) that are produced and secreted by plasma cells in response to specific antigens.

- The majority of cells in the clone become **plasma cells**, which produce and then release large quantities of their specific antibody.
- A few B cells in the clone do not develop into plasma cells but remain as inactive **memory cells**.

> **Plasma cells** are antibody-secreting cells produced by the division of activated B cells.

17 During clonal selection, B cells differentiate into plasma cells. Give *three* ways in which you expect the ultrastructure of plasma cells to be adapted to their role in secreting antibodies.

Answer on p. 201

> **Typical mistake**
>
> When writing about the complementary fit of a cell surface receptor and an antigen, students often use the term 'active site'. Remember, only enzyme molecules have active sites.

The nature of antibodies

Antibodies are immunoglobulins (abbreviated to Ig) released by plasma cells. Figure 6.8(a) shows the structure of the most common antibody, called IgG. It has:
- four polypeptides linked by disulfide bonds (−S−S−)
- two large polypeptides (heavy chains) and two small polypeptides (light chains)
- sites at the very ends of the light chains that bind with a specific antigen

Figure 6.8(b) shows how these antibodies can bind to antigens on a number of foreign cells, causing clumping of the cells. This clumping (called **agglutination**) aids destruction of the foreign cells.

> **Revision activity**
>
> Adapt the diagram in Figure 6.7 to represent the sensitisation of a B cell and the production of plasma cells and memory cells.

> **Agglutination** refers to the clumping together of pathogenic cells by antibodies.

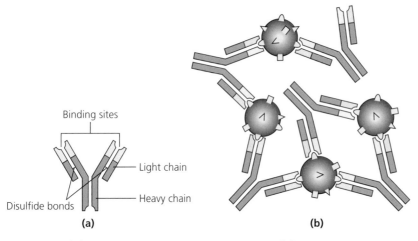

(a) Binding sites, Light chain, Disulfide bonds, Heavy chain **(b)**

Figure 6.8 (a) The structure of an IgG antibody. (b) Several IgG antibodies clump bacteria carrying the same surface antigen

> **Typical mistake**
>
> Students regularly confuse terms with similar components. Take care to use the terms antibiotic, antibody and antigen correctly.

Types of humoral immunity

Humoral immunity can develop in a number of ways:
- **Active immunity** is produced by antibodies in response to exposure to antigens.
  - Natural active immunity develops after exposure to antigens in the environment.
  - Artificial active immunity develops after the introduction of antigens, for example by vaccination.
- **Passive immunity** is produced by antibodies transferred from another organism.
  - Natural passive immunity is conferred by the transfer of maternal antibodies across the placenta or in breast milk.
  - Artificial passive immunity is conferred by the injection of antibodies.

Now test yourself

TESTED

18 Suggest *one* instance when it might be advantageous to use artificial passive immunity in medicine.
19 Suggest the importance of having both the cell-mediated immune response and the humoral immune response.

Answers on p. 201

The role of memory cells

Memory cells are inactive cells produced during the clonal selection of B cells and of T cells. Their role is to produce a faster and stronger **secondary immune response** to any further infection by the specific pathogen. They:

- remain in the bloodstream
- bind to their specific antigen should a second infection occur
- rapidly undergo clonal selection

You can compare the response to first exposure to an antigen (the primary response) with the secondary response in Figure 6.9.

> The **secondary immune response** is the response to infection by a pathogen carrying antigens to which there are memory cells already present in host's body.

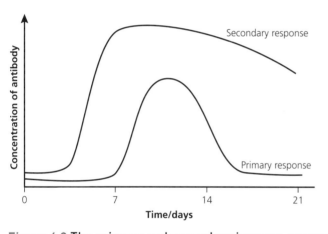

Figure 6.9 The primary and secondary immune responses

Vaccination and immunisation programmes

REVISED

An immunisation programme is undertaken to stimulate a primary response to a pathogen among the people in a population before they encounter the pathogen naturally. The primary response is stimulated using specific **vaccines**.

To make a vaccine, antigens from a pathogen are prepared in such a way that, when injected into a recipient, the antigen will result in the production of specific antibodies but will not cause disease. The vaccinated person develops artificial active immunity without suffering the disease.

> **Vaccines** are relatively harmless preparations of antigens derived from a specific pathogen, which result in the production of antibodies by a vaccinated recipient.

Now test yourself

TESTED

20 Suggest *two* reasons why it might not be safe to vaccinate everyone in a community.

Answer on p. 201

Often, it is not possible to vaccinate everyone within a population. This might not be a problem, however, as long as a sufficiently large percentage of the population is protected through vaccination to achieve **herd immunity**. Factors that prevent the development of herd immunity include the following:

- Communities with objections to vaccination. For example, the Christian Science church rejects many forms of medical care, including vaccination.
- Communities receiving poor advice that a vaccine is harmful. For example, an erroneous scare that the MMR vaccine could lead to autism resulted in the 2011 measles outbreak in the UK.

> **Herd immunity** arises when a sufficiently high percentage of the population has been vaccinated against a pathogen to make it unlikely that a susceptible person will be infected.

> **Revision activity**
>
> Construct a table to compare and contrast the action of T cells and B cells.

Exam practice

1 (a) The following events occur during the humoral immune response. Place the numbers 1 to 6 in the right-hand column to indicate the order in which they occur. Use 1 to indicate the earliest event and 6 to indicate the latest event. [3]

Event	Order (earliest = 1)
B cell sensitised	
Plasma cells formed and release antibodies	
T cell activated when it encounters antigen-presenting cell with peptide fragments that complement its own surface receptors	
B cell activated by cytokines	
Macrophage ingests foreign cell and becomes antigen-presenting cell, displaying peptide fragments from foreign cell on its surface membrane	
T cell divides, producing T helper cells that secrete cytokines	

(b) What is the role of memory cells. [3]

2 A group of ecologists wanted to find out if the ground beetles they were studying ate slugs. Since slugs have no hard body parts, the ecologists could not analyse the beetles' gut contents using a microscope to detect the remains of slugs. Instead, they used an enzyme-linked immunosorbent assay (ELISA) technique to test for the presence of any slug proteins in the beetles' guts. The diagram summarises what they did.

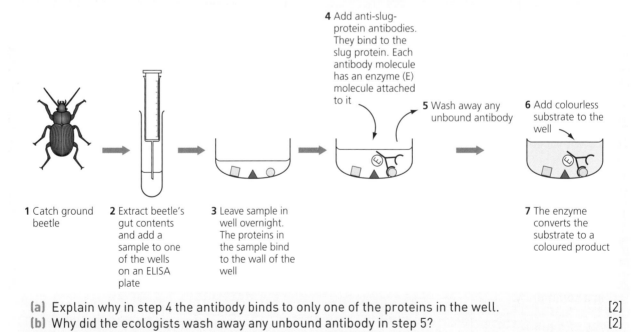

4 Add anti-slug-protein antibodies. They bind to the slug protein. Each antibody molecule has an enzyme (E) molecule attached to it

5 Wash away any unbound antibody

6 Add colourless substrate to the well

1 Catch ground beetle

2 Extract beetle's gut contents and add a sample to one of the wells on an ELISA plate

3 Leave sample in well overnight. The proteins in the sample bind to the wall of the well

7 The enzyme converts the substrate to a coloured product

(a) Explain why in step 4 the antibody binds to only one of the proteins in the well. [2]
(b) Why did the ecologists wash away any unbound antibody in step 5? [2]
(c) Suggest how the ecologists obtained antibodies against the proteins present in slugs (the anti-slug-protein antibodies in the diagram). [3]

3 A microbiologist intended to use a colorimeter to measure the number of cells in bacterial cultures. First, she used a haemocytometer slide to count the number of cells in a serial dilution of a bacterial culture. She used these counts to produce a calibration curve.

(a) Give *two* aseptic techniques she would use when producing a serial dilution of the bacterial culture. [2]

(b) Explain why she needed to produce a calibration curve. [1]

(c) In a single 0.05 mm × 0.05 mm cell of a haemocytometer, she counted 14 bacterial cells. The depth of the cell was 0.1 mm and the dilution of the sample was 10^{-6}. Calculate the density of cells in the undiluted culture. Show your working. [2]

4 (a) Give *two* ways in which bacteria act as agents of infection. [2]

(b) Explain why antibiotics can be effective against bacteria but not against viruses. [2]

In 2015/16, a number of women in South America gave birth to babies with microcephaly. This was thought to be the result of these women being infected by the Zika virus while they were pregnant. The Zika virus is spread by the bite of female *Aedes aegypti* mosquitoes. In adults, infection by the Zika virus usually produces mild symptoms, treated by rest. Microcephaly severely limits the growth and development of the brain of affected infants.

(c) Suggest why the focus of disease control should be on preventing spread of the Zika virus. [3]

Answers and quick quiz 6 online

ONLINE

Exam tip

Question 4 tests your recall and understanding of other topics from the specification.

Summary

Microbial techniques

- Scientists use aseptic techniques when culturing microorganisms.
- Microorganisms can be cultured in broth cultures, on agar plates and in selective media.
- The growth of a microbial culture can be measured using direct cell counts, dilution plating, change in mass and by optical methods.
- The typical growth of bacteria in batch culture includes a lag phase, log phase, stationary phase and death phase. The exponential growth rate constant is the rate of growth during the log phase.

Pathogens and antibiotics

- Bacteria can act as pathogens by invading and destroying host tissues (e.g. *Mycobacterium tuberculosis*), by producing endotoxins (e.g. *Salmonella* spp.) and by producing exotoxins (e.g. *Staphylococcus* spp.).
- Other groups of organisms can also act as pathogens, including viruses (e.g. influenza virus), fungi (e.g. stem rust fungus) and protoctists (e.g. *Plasmodium* spp.).
- Bacteriostatic antibiotics slow, and bactericidal antibiotics stop, the metabolism of bacteria.
- As a result of natural selection, many species of bacteria have become resistant to several antibiotics. Methods to control the spread of antibiotics are not without their difficulties.

- There are social, economic and ethical implications associated with methods for controlling endemic diseases, such as malaria.

Response to infection

- Macrophages and neutrophils are phagocytic cells that protect the body by engulfing and destroying pathogens.
- There are two types of lymphocyte: T cells and B cells. Both have surface receptors that are complementary to an antigen present on the surface of a pathogen.
- When presented with its complementary antigen, a lymphocyte is sensitised and divides repeatedly by mitosis to produce activated lymphocytes and inactive memory cells (clonal selection).
- Lymphocytes produce active immunity. T cells produce the cell-mediated immune response and B cells produce the humoral immune response.
- Vaccination involves the administration of a non-disease-causing form of the pathogen, stimulating the active immune response by lymphocytes. So long as a large proportion of a community has been immunised, there is little risk of unvaccinated people being infected (the principle of herd immunity).

7 Modern genetics

This topic builds on content from Topic 1 (DNA and protein synthesis) and Topic 3 (DNA sequencing can be used to distinguish between species).

Using gene sequencing

Gene sequencing involves finding the nucleotide base sequence of a cell's **genome**. The amount of DNA obtained from cells is very small. In order to analyse it, the DNA must be **amplified**, i.e. copied many times.

> A **genome** is all the genetic material in a single somatic cell from an organism. In your cells, this includes the coding and non-coding DNA, but not the RNA.
>
> A DNA molecule is **amplified** when a large number of identical copies are made from it.

Now test yourself

TESTED ☐

1 Name *three* sources of DNA in a plant cell.

Answer on p. 202

The polymerase chain reaction (PCR) amplifies DNA samples

REVISED ☐

The **polymerase chain reaction** is an important tool for amplifying small samples of DNA to provide enough for analysis. The process is similar to the semi-conservative replication of DNA that you learned about in Topic 1.

The PCR process is carried out in automated thermal cyclers that are programmed to change the temperature (Figure 7.1). Within the thermal recycler is:
- the DNA to be amplified (the template DNA)
- DNA nucleotides
- a thermostable **DNA polymerase** — the enzyme that catalyses the formation of phosphodiester bonds between DNA nucleotides
- **primers**, which:
  ○ are short sections of single-stranded nucleic acid (oligonucleotides)
  ○ have a nucleotide base sequence that is complementary to a region of DNA close to, or at the start of, the DNA sequence to be amplified
  ○ enable DNA polymerase molecules to attach to template strands and start the replication process

The PCR is normally used to amplify parts of DNA molecules, rather than whole DNA molecules. Consequently, the products of PCR are fragments of DNA.

> The **polymerase chain reaction** (PCR) is a process used to make large amounts of identical DNA from a small sample.
>
> **Primers** are single-stranded oligonucleotides that allow DNA polymerase to begin replication of a template strand of DNA.

> **Typical mistake**
>
> Examiners are regularly told that DNA polymerase pairs nucleotides with complementary bases. It does not, it catalyses the formation of phosphodiester bonds between nucleotides that have already formed base pairs.

Now test yourself

TESTED ☐

2 Why must the DNA polymerase used in the PCR be thermostable?
3 The 'nucleotides' used in the PCR are actually deoxynucleoside triphosphates, such as dATP, which are hydrolysed during the formation of a phosphodiester bond. Use your knowledge of ATP to suggest *one* advantage of using deoxynucleoside triphosphates, rather than nucleotides.

Answers on p. 202

Separating DNA fragments using electrophoresis

REVISED

The phosphate ions within DNA are anions, i.e. they have a negative charge. This means that if DNA fragments are placed in an electric field, they move towards the positive pole. This is the basis of **electrophoresis**. You can see in Figure 7.2 one way in which electrophoresis is used to separate DNA fragments:

- Samples of DNA are placed in wells at one end of an agarose gel.
- A power supply creates an electric field along the gel, with the wells at the negative pole.
- The negatively charged fragments migrate from the negative to the positive pole.
- Pores in the gel allow smaller fragments to move faster than larger fragments.
- The gel is fragile, so an imprint is made onto a more robust nylon membrane.
- **Probes** — short single strands of DNA, labelled with radioactive phosphorus (^{32}P) — are used to locate the bands on the nylon membrane.
- An X-ray film is used as a permanent record of the results of electrophoresis.

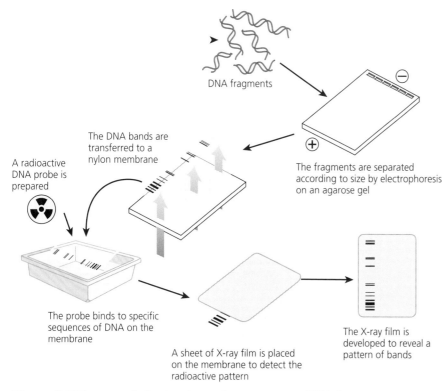

Figure 7.2 The use of electrophoresis to separate DNA fragments

DNA fragments

The DNA bands are transferred to a nylon membrane

A radioactive DNA probe is prepared

The probe binds to specific sequences of DNA on the membrane

A sheet of X-ray film is placed on the membrane to detect the radioactive pattern

The fragments are separated according to size by electrophoresis on an agarose gel

The X-ray film is developed to reveal a pattern of bands

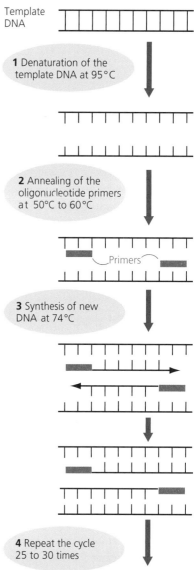

Template DNA

1 Denaturation of the template DNA at 95°C

2 Annealing of the oligonucleotide primers at 50°C to 60°C

Primers

3 Synthesis of new DNA at 74°C

4 Repeat the cycle 25 to 30 times

Figure 7.1 The key steps in the PCR

Electrophoresis is used to separate charged substances by causing them to migrate through a porous medium placed in an electric field.

Probes are short, single-stranded chains of nucleotides with radioactive phosphorus (^{32}P) or fluorescent markers that will pair with complementary base sequences within a target DNA fragment.

Typical mistake

Students often confuse the structure and role of primers and probes. Make sure that you are clear what each does.

Now test yourself

TESTED

4 Which of the DNA fragments in Figure 7.2 is the smallest? Explain your answer.

Answer on p. 202

Using the PCR and electrophoresis in gene sequencing

The standard way to sequence DNA is to use a **chain termination** process. This process was outlined in Topic 3 (page 58).

- The chain termination process uses the PCR, except that it includes the use of slightly abnormal nucleotides, called **dideoxynucleotides** (or ddN).
- The deoxyribose residue of these nucleotides has a hydrogen atom instead of a hydroxyl group on carbon atom 3 that prevents it forming a phosphodiester bond.
- When one of these dideoxynucleotides is randomly incorporated into a developing DNA chain, the action of DNA polymerase is stopped.

To use the chain termination sequence technique with standard gel electrophoresis, you would use four separate tubes. Into each, you would add:

- copies of the single-stranded DNA to be sequenced (the template DNA)
- DNA polymerase
- primers
- molecules of all four normal DNA nucleotides
- molecules of only one type of dideoxynucleotide carrying adenine (ddNA) or cytosine (ddNC) or guanine (ddNG) or thymine (ddNT)

Once in the thermal cycler, the PCR occurs in each tube. When, by chance, a dideoxynucleotide is added to a developing DNA strand, replication stops. This results in a mixture of DNA strands of different lengths in each tube.

After one replication using the PCR, you can place the contents of each tube into separate wells of a gel and use electrophoresis to separate the fragments. Figure 7.3 shows how the base sequence of the DNA can be elicited using this technique.

Exam tip

The specification does not give details on the method of gene sequencing, so it would not be valid for you to be asked to recall one.

Revision activity

Use information in the text to sketch a diagram to represent the structure of a dideoxyribose molecule.

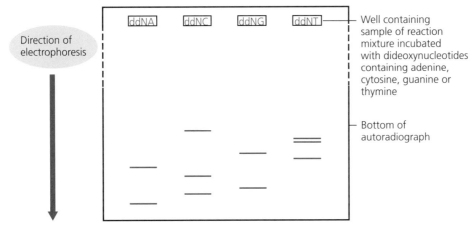

Figure 7.3 Separation of DNA fragments during the DNA chain termination process

- The fragment that has moved furthest is in the lane with the contents of the tube containing dideoxynucleotides carrying adenine (ddNA). This tells us that adenine is the first base in the complementary strand, so thymine must be the first base in the template strand.

- The fragment that has moved next furthest is in the lane with the contents of the tube containing dideoxynucleotides carrying cytosine (ddNC). This tells us that the second base in the complementary strand is cytosine, so guanine must be the second base in the template strand.
- By continuing along the sequence of fragments, you can work out the complete base sequence of the complementary strand being constructed by the PCR and, from that, the base sequence of the single-stranded template DNA that was placed in each tube.

Now test yourself

TESTED ☐

5 Which component of DNA causes it to migrate during electrophoresis?
6 Use information in Figure 7.3 to give the full base sequence of:
  (a) the complementary DNA strand
  (b) the template DNA strand

Answers on p. 202

Gene sequencing can be automated

In an **automated gene sequencer**:
- only one reaction tube is used
- molecules of the four types of dideoxynucleotides are labelled with a different dye
- electrophoresis occurs in gel within a capillary tube
- the coloured bands in the capillary tubing are 'read' and represented as a sequence of coloured peaks, as in Figure 7.4
- the base sequences of the different fragments of DNA are pieced together to give the base sequence of the template DNA

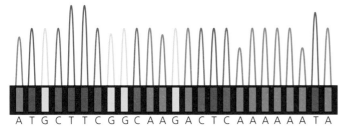

Figure 7.4 Part of the printout from an automated gene-sequencing machine. The dideoxynucleotides have been labelled with different dyes: ddNA green, ddNC blue, ddNG yellow and ddNT red

Now test yourself

TESTED ☐

7 What is represented by ddNA?

Answer on p. 202

Turning fragment sequences into whole genome sequences

The chain termination technique results in a large number of fragments of DNA for which the base sequences are known. This is rather like having all the pieces of a jigsaw puzzle without having the picture of the finished puzzle.

To complete the sequencing for an entire DNA molecule, you must:
- find overlaps in the base sequences of the DNA fragments
- arrange these overlaps to find which fragment overruns an adjacent fragment
- arrange all the overlapping fragments into a consensus sequence, i.e. one that all scientists in the field agree is an authentic representation of the sequence of each DNA molecule in the genome

Thankfully, all this is done by computer.

Using the results of whole genome sequencing

Once known, the base sequence of DNA can be used to find the number, and base sequence, of all the genes in the genome. You can then use these gene sequences to:
- predict the amino acid sequence of the polypeptides encoded by each gene
- find the base sequence of the allele of a gene that results in a medical condition
- find similarities in the base sequences of common genes to work out evolutionary relationships between different organisms (page 59)

> **Exam tip**
>
> Don't compartmentalise your revision. Examiners could easily set a question that combines this topic with understanding of content from Topic 1 (nucleic acids and protein synthesis) and Topic 3 (distinguishing different species).

TESTED

Now test yourself

8 How could scientists use the results of gene sequencing to predict the amino acid sequence of a polypeptide?

Answer on p. 202

Using the PCR and electrophoresis in DNA profiling

REVISED

In eukaryotes, much of the DNA does not encode amino acids. These **non-coding sections** occur within genes — **introns** — and between genes, including **short tandem repeats** (**STRs**).

We all inherit a unique combination of these short tandem repeats, half on chromosomes from our mother and half on chromosomes from our father. This is the basis of **DNA profiling**.

> **Short tandem repeats** are sections of DNA with a repeated sequence of nucleotide bases, for example TATATA.

In DNA profiling:
- the PCR is used to amplify short sections of DNA that are known to contain STRs from, say, a couple and their child
- gel electrophoresis is used to separate the DNA fragments containing these STRs from each person
- probes are used to identify DNA fragments containing the target short tandem repeats
- the positions of the DNA fragments from each person are compared

Figure 7.5 shows DNA profiles of two parents and their son. You can see that each STR-containing DNA band from the son is matched by one from his mother or his father.

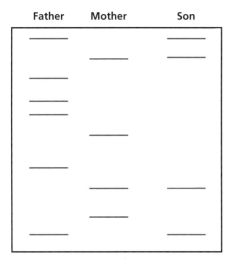

Figure 7.5 **DNA profiles of a couple and their son**

Using the results of DNA profiling

DNA profiles have a variety of uses. For example, they can be used to:
- establish parentage, for example when members of a family of refugees have been separated and are trying to be re-united
- test paternity, for example when a man denies, or has reason to doubt, that he is the father of a child
- identify criminals, for example by comparing the DNA profiles of suspects with that of DNA left at the scene of a crime

Now test yourself   TESTED

9 Blood is often left at the scene of a crime. What is the likely source of DNA in a blood sample?

Answer on p. 202

Factors affecting gene expression

In a multicellular organism, some genes are **expressed** in every cell but many are not. For example, in your body the gene encoding the enzyme alcohol dehydrogenase is expressed only in your liver cells.

Expression of a gene can be regulated by stimulating or repressing:
- transcription — the production of mRNA from DNA
- translation — the production of a polypeptide using the base sequence of mRNA

> A gene is said to be **expressed** if its base sequence is transcribed into mRNA that is then translated by ribosomes.

As you know from Topic 1 (Figure 1.14, page 18), for a gene to be transcribed RNA polymerase must be able to attach to it. Several factors affect the ability of RNA polymerase to attach to a gene and by doing so they regulate transcription.

Transcription factors regulate transcription   REVISED

In eukaryotes, transcription of a gene is controlled by a **promoter**. This is a region of DNA close to, and upstream from, the gene (Figure 7.6). To initiate transcription of a gene:

> A **promoter** is a region of DNA upstream from the gene whose transcription it regulates.

- one or more proteins, called **transcription factors**, bind to the DNA of the gene's promoter
- RNA polymerase then binds to the DNA–transcription factor complex
- as a result, the RNA polymerase is activated and moves to the start of the gene
- the RNA polymerase transcribes the antisense strand of the gene

So, unless the relevant transcription factors are present, RNA polymerase cannot transcribe a particular gene.

> **Transcription factors** are proteins that bind to regions of DNA, called promoters, that regulate gene expression.

Now test yourself

TESTED

10 Explain the term 'upstream of a gene'.

Answer on p. 202

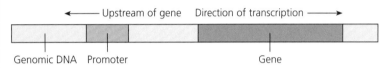

Figure 7.6 In eukaryotes, each gene is downstream from its promoter. RNA polymerase is able to transcribe a gene only when several transcription factors combine at the promoter

Now test yourself

TESTED

11 Use your understanding of biological principles to suggest what enables RNA polymerase to bind to the DNA–transcription factor complex at a gene's promoter.

Answer on p. 202

Since many genes are also regulated by other genes that are, in turn, regulated by transcription factors, a single gene might be regulated by a whole cascade of transcription factors.

Post-transcriptional modification of mRNA

REVISED

Look back to Figure 1.15 (page 18) to remind yourself of how pre-mRNA is modified in eukaryotic cells by removal of the introns. In fact, once introns are removed, eukaryotic cells can splice the exons back together in many different ways. The result is that many different mRNA sequences can be produced from one gene. This is demonstrated in Figure 7.7 using a gene that, depending on which of its exons are spliced together, encodes calcitonin or calcitonin gene-related peptide (CGRP).

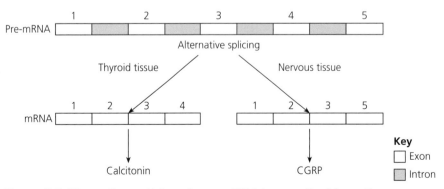

Figure 7.7 Alternative splicing of pre-mRNA transcribed from the calcitonin gene in humans results in tissue-specific products

Epigenetic modification regulates transcription

You saw in Topic 1 (page 20) how changes to the DNA base sequence of a gene — gene mutations — can affect the expression of a gene. Inheritable changes in gene function can also occur without changes in the DNA base sequences. These result from epigenetic changes.

Epigenetics involves changes in gene function that are:
- not caused by a change in the nucleotide base sequence of the gene involved
- passed on from generation to generation
- important in ensuring that cells in your body differentiated into different tissue types

> **Epigenetics** involves a heritable change in the expression of a eukaryotic gene without any change in its sequence of nucleotide bases.

Two epigenetic modifications that regulate transcription are:
- histone acetylation
- DNA methylation

Histone acetylation allows genes to be transcribed

DNA is normally wound tightly around molecules of a protein, called **histone**. You can see in Figure 7.8(a) that parts of the histone molecules protrude as histone 'tails'. The amino groups of amino acids in these 'tails' (especially the amino acid lysine) can be modified by addition or removal of acetyl groups as shown in the formula:

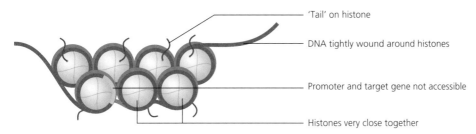

(a) Histones not acetylated

(a) Histones not acetylated

- 'Tail' on histone
- DNA tightly wound around histones
- Promoter and target gene not accessible
- Histones very close together

(b) Histones acetylated

(b) Histones acetylated

- Acetyl groups on histone tails
- Histones no longer close together
- Promoter and target gene now accessible to transcription factors and RNA polymerase

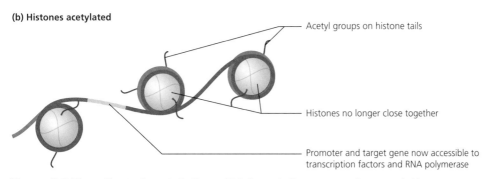

Figure 7.8 The effect of acetylation of histone tails on gene transcription

- Figure 7.8(a) shows that when the histone 'tails' are *not* acetylated, the DNA is tightly packed. In this condition, transcription factors and RNA polymerase cannot get to the genes or their promoters in that region. Those genes cannot be transcribed.
- Figure 7.8(b) shows that when the histone 'tails' *are* acetylated, the DNA is loosely packed. In this condition, transcription factors and RNA polymerase can get to the genes and their promoters in that region. Those genes can be transcribed.

Now test yourself

TESTED ☐

12   Name *one* other process in cells that involves acetylcoenzyme A.

Answer on p. 202

DNA methylation prevents genes being transcribed

In mammals, **DNA methylation** involves the cytosine in a cytosine-carrying nucleotide that is next to a guanine-carrying nucleotide (represented by **CpG**). Figure 7.9 shows the result of methylation of cytosine, catalysed by the enzyme **DNA methyltransferase**.

> **DNA methylation** involves the addition of a methyl group (CH_3) to a DNA nucleotide.

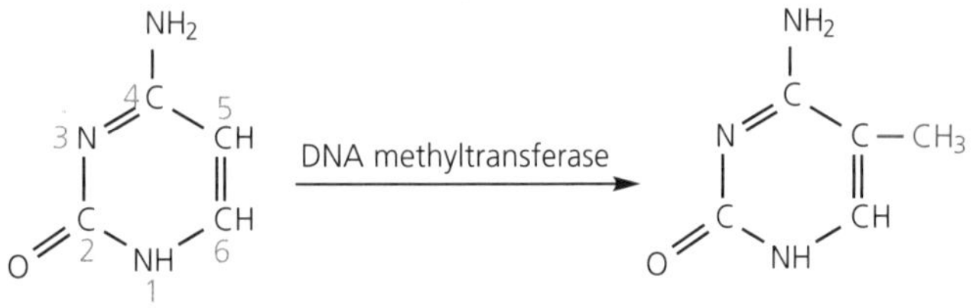

Figure 7.9 Methylation of cytosine

The presence of methylated CpG sequences near promoters:
- prevents the activation of RNA polymerase described above
- stops transcription of the downstream gene ('silences' the gene)

Patterns of methylation are inherited because:
- DNA replication produces new DNA molecules in which only the template strand is methylated
- DNA methyltransferases methylate the new unmethylated strand to complement the pattern on the methylated template strand

Now test yourself

TESTED ☐

13   Which part of a chromosome is affected by:
   (a) acetylation?
   (b) methylation?

Answer on p. 202

Non-coding RNA regulates translation

REVISED

Animal and plant cells produce large numbers of RNA molecules that, unlike mRNA and tRNA, do not carry a code for amino acids. They are called **non-coding RNA**.

Now test yourself   TESTED

14 What is the function of tRNA?

Answer on p. 202

MicroRNA (miRNA) is one type of non-coding RNA. It prevents the translation of mRNA.

- miRNAs are themselves encoded by genes. A single miRNA molecule transcribed from its gene is a fairly long, double-stranded RNA molecule, referred to as pri-miRNA (primary miRNA).
- Each pri-miRNA molecule is hydrolysed within the nucleus to form short, single-stranded miRNA molecules that are about 20 nucleotides long.
- In the cytoplasm, these single-stranded miRNA molecules bind to base sequences within mRNA molecules that are complementary to their own. Figure 7.10 shows what then happens.
- If the base pairing is incomplete, as in Figure 7.10(a), the miRNA interferes with the activity of the ribosome, stopping translation.
- If the base pairing is perfect, as in Figure 7.10(b), the miRNA attracts enzymes that hydrolyse the mRNA.

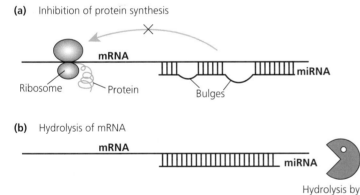

(a) Inhibition of protein synthesis

mRNA

Ribosome   Protein   Bulges   miRNA

(b) Hydrolysis of mRNA

mRNA

Perfect RNA–RNA base-pairing   miRNA   Hydrolysis by enzyme

Figure 7.10 miRNA prevents translation of mRNA

Stem cells

The nature of stem cells

REVISED

In mature organisms, cells have differentiated, becoming specialised for particular functions. During the process of differentiation, most of these cells lose the ability to divide by mitosis. The cells that retain the ability to divide by mitosis are called **stem cells**.

> A **stem cell** is a type of cell that retains the ability to divide by mitosis.

Figure 7.11 shows how stem cells are able to divide to produce:
- new stem cells
- cells that differentiate into specialised cells

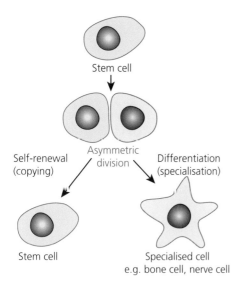

Figure 7.11 Division of stem cells results in self-renewal as well as in specialised cells

Table 7.1 summarises the three types of stem cell found in humans.

Table 7.1 Three types of human stem cell

Type of stem cell	Properties of stem cell	Location in humans
Totipotent stem cell	Can divide to replicate itself and produce any cell in the body and in the placenta	Zygote
Pluripotent stem cell	Can divide to replicate itself and produce any cell in the body but not cells of the placenta	Blastomeres (look back to Figure 2.12, page 43, to remind yourself of these cells)
Multipotent stem cell	Can divide to replicate itself and produce a few cell types	Found in various locations in the body, for example: ● haematopoietic cells in bone marrow produce blood cells ● mesenchymal cells in bone marrow produce bone cells and cartilage cells

Now test yourself

TESTED

15 During human development, a totipotent zygote produces pluripotent blastomeres that produce multipotent adult stem cells that produce fully differentiated somatic cells. What controls this process?

Answer on p. 202

The clinical use of stem cells

REVISED

Many clinical conditions are caused by a change in the function, or death, of tissues. Stem cells could be used to replace these cells with healthy, new ones. For example:

● A tissue containing abnormal cells could be destroyed or removed and replaced with tissue containing stem cells.
● Dead tissue could be replaced with tissue containing stem cells.
● Given an appropriate scaffolding and relevant transcription factors, stem cells could be used to produce new tissues or organs in the laboratory, ready for transplant.

Using embryonic stem cells

Embryonic stem cells are blastomeres. Since they are pluripotent, they would be ideal for the types of clinical uses outlined above.

- During *in vitro* fertilisation (IVF), many blastocysts are produced from the eggs and sperm of a couple being helped to conceive.
- Few of these are ever used, leaving 'spare' blastocysts.
- With the consent of both members of the couple, these 'spare' blastocysts could be used as a source of pluripotent stem cells for clinical use.
- The above process raises many legal, social and ethical issues that limit its availability throughout the world.

Induced pluripotent stem cells (iPS cells)

An alternative to the use of embryonic stem cells has been developed that does not raise the ethical issues associated with removing stem cells from blastocysts.

Fibroblasts are found in connective tissue. They can be treated in such a way that their differentiation process is reversed and they regain the properties of pluripotent stem cells.

- In laboratory cultures that include four specific transcription factors, fibroblasts can be induced to become pluripotent stem cells.
- If made from a patient's own fibroblasts, injection of these induced pluripotent stem cells would avoid rejection by the patient's immune system.
- Unfortunately, the yield of iPS cells in laboratory cultures is increased by activating proto-oncogenes in the fibroblast cells, which raises safety concerns.

Now test yourself

16 Why would use of iPS cells induced from a patient's own fibroblasts avoid rejection by the patient's immune system?
17 Why would the activation of proto-oncogenes raise safety concerns?

Answers on p. 202

Gene technology

Recombinant DNA technology

REVISED

Recombinant DNA technology involves:
- identifying the gene controlling a particular characteristic in one organism
- producing multiple copies of the gene
- inserting copies of the gene into the DNA of another organism

> **Gene technology** is the term used to describe techniques that can be used to alter an organism's genome.

Producing recombinant DNA

Providing that the nucleotide base sequence at the start of a gene is known, the PCR can be used to amplify the gene, as described on page 128.

Before copies of this gene can be inserted into the genome of another organism, they must be inserted into a **vector**. Bacterial plasmids are commonly used vectors (look back to Figure 2.1, page 30, to remind yourself of plasmids).

> In DNA technology, a **vector** is a structure that is used to carry DNA into another cell.

- A bacterial plasmid is opened using an enzyme called a **restriction endonuclease**.
- The gene of interest is spliced into the opened plasmid, using an enzyme called **DNA ligase** to form the phosphodiester bonds between the plasmid and the gene. This is called **annealing**.
- The result is a plasmid containing a foreign gene, as shown in Figure 7.12.

> **Exam tip**
>
> The term 'vector' has several different meanings in science and mathematics. If asked, make sure that you answer in the context of the question.

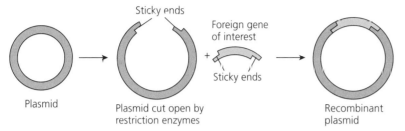

Figure 7.12 Production of recombinant DNA using a bacterial plasmid as a vector

Now test yourself

TESTED ☐

18 Explain why the final plasmid in Figure 7.12 is described as 'recombinant DNA'.

Answer on p. 202

The action of restriction endonucleases

Restriction endonucleases have active sites that are complementary to specific DNA base sequences, called **recognition sequences**. Figure 7.13 shows the recognition sequences of two different restriction endonucleases.

> **Recognition sequences** are specific palindromic sequences of nucleotide bases that are the targets of restriction endonucleases.

a) Production of blunt ends by the restriction enzyme Alu1

b) Production of sticky ends by the restriction enzyme EcoR1

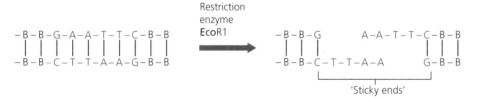

Figure 7.13 The action of two restriction endonucleases. The recognition sequences are shown in red

- Alu1 hydrolyses DNA, leaving no unpaired bases at the ends of the strands. These are referred to as 'blunt ends'.
- EcoR1 hydrolyses DNA, leaving unpaired bases at the ends of the strands. These are referred to as 'sticky ends'.
- Using DNA ligase to join a piece of DNA into an opened plasmid is more successful if both have complementary 'sticky ends', as shown in Figure 7.12.

- If the target gene has been removed from a DNA molecule, it should be removed with the same restriction endonuclease as that used to open the plasmid.
- If the gene has been amplified using the PCR, 'sticky ends' must be added that are complementary to the ends of the opened plasmid.

TESTED

Now test yourself

19 Use Figure 7.13 to explain the term 'palindromic base sequence'.
20 If a gene is removed from a DNA molecule, why must the same restriction endonuclease be used to open the plasmid that will act as a vector?

Answers on p. 202

Inserting recombinant DNA into target cells and finding transformed cells

Many species of bacteria are able to take up plasmids. One method to encourage bacteria to take up recombinant plasmids is to:
- soak the bacteria and plasmids in ice-cold calcium chloride solution
- apply a brief heat 'shock' of 42°C for 2 minutes

Identifying transformed bacteria using antibiotic resistance marker genes and replica plating

Using recombinant plasmids to transform bacteria is a 'hit-and-miss' process. Not all the bacteria will have taken up a plasmid and, even if they have, many of those plasmids will have annealed without having combined with the gene of interest. Scientists need a method to identify bacterial cells that have taken up the gene of interest (i.e. are **transformed bacteria**).

Using **antibiotic resistance marker genes** and **replica plating** is one of the earliest methods used to identify transformed bacteria. In the following example, bacteria are transformed using, as a vector, a plasmid that contains genes conferring resistance to two antibiotics.
- Figure 7.14(a) shows the plasmid with genes conferring resistance to two antibiotics, which is used as a vector.
- Figure 7.14(b) shows one of these plasmids that has also taken up the gene of interest in the correct position.

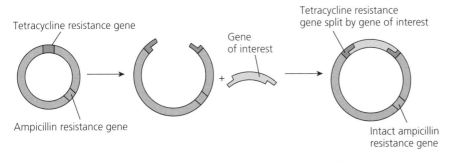

(a) Normal plasmid   (b) Plasmid with recombinant DNA

Figure 7.14 A plasmid with two marker genes

Figure 7.15 shows how these marker genes enable transformed bacteria to be identified.
- Following treatment and heat-shock, samples of the bacteria are grown on agar plates.

- Figure 7.15 shows the results of plating one sample of bacteria. The upper plate has six colonies but which, if any, contains transformed bacteria?
- To find out, an imprint of the colonies is taken, by pressing a nylon sheet gently onto the plate, and transferring samples of the colonies to a second plate with agar containing the antibiotic ampicillin. A second imprint is pressed onto a third plate with agar containing the antibiotic tetracycline.
- The lower plates in Figure 7.15 show the results after incubation.
- Of the six bacteria that formed colonies in the upper plate, only four (1, 2, 4 and 6) had taken up the plasmids as shown by their ability to grow in the presence of ampicillin.
- Of these, two (2 and 4) were unable to grow on agar containing tetracycline. These must be the bacteria that took up the plasmids that contained the gene of interest.

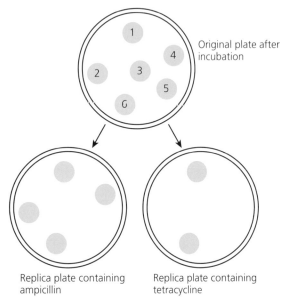

Original plate after incubation

Replica plate containing ampicillin

Replica plate containing tetracycline

Figure 7.15 The results of replica plating of bacterial cultures growing on nutrient agar onto one agar plate containing ampicillin and another agar plate containing tetracycline

Now test yourself

TESTED

21 How can you conclude from Figure 7.15 that the cells that formed colonies 1, 2, 4 and 6 had each taken up a plasmid?

22 How can you conclude from Figure 7.15 that the plasmid taken up by the cells that formed colonies 1 and 6 did *not* contain the gene of interest?

Answers on p. 202

Other vectors

Vectors other than plasmids can be used. They include:
- **viruses** — during latency (page 35), the nucleic acid of a virus has become inserted into the DNA of the host cell. If the virus has been engineered to contain the target gene, that gene will become part of the host cell's DNA.
- **gene guns** — laboratory machines that use air pressure to fire particles of heavy metals coated with the target DNA into cells

Uses of recombinant DNA technology

An early use of recombinant DNA technology was to insert genes for useful products into bacteria. When grown in continuous culture (page 112), these bacteria produce the useful product, which can then be harvested. An example of this use of recombinant DNA technology is the production of human insulin for use by sufferers of type 1 diabetes.

Knockout mice

Traditionally **knockout mice** are made using embryonic stem cells:
- Researchers change the DNA sequence of a gene to produce an inactivated mutant gene.
- They inject this mutant gene into mouse embryos and hope it is incorporated into the embryonic DNA.
- Some of the resulting adult mice will have bodies with cells that contain the mutant gene and cells that contain the normal gene. From these, the researchers find the male mice that produce sperm containing the mutant gene.
- They breed these males with normal females, hoping that some offspring will contain one normal gene and one mutant gene in every cell, i.e. are heterozygous for this gene.
- By breeding these heterozygotes together, they obtain some mice with two copies of the mutant gene, i.e. are homozygous for the mutant gene. These are the knockout mice.

> **Knockout mice** are homozygous for a gene that has been inactivated by DNA technology. They are commonly used for biomedical research.

TESTED

Now test yourself

23 How could scientists find the function of a gene that has been inactivated?
24 Suggest how scientists could use knockout mice to test a therapy to control a disease.

Answers on p. 202

Genetically modified (GM) soya beans

Beans of the soya plant (*Glycine max*) are used around the world as a food for humans and livestock. Using recombinant DNA technology, GM soya plants have been produced that:
- are resistant to a commonly used weed killer (glyphosate)
- are resistant to common fungal pests
- are resistant to drought
- produce active pharmacological ingredients
- have a low concentration of linoleic acid — a fatty acid that becomes oxidised to harmful trans fats when cooked

The debate about recombinant DNA technology

Recombinant DNA technology allows the development of new varieties of organism much faster than the traditional breeding and artificial selection process that has developed our crops and livestock. Other advantages of the technique have been given above. Public concern relates to potential disadvantages, which are outlined below:
- A transferred gene might interact with the normal genome of the recipient in an unknown and harmful way.
- GM foods might be harmful to human health.
- GM crops might be harmful to beneficial insects, such as bees.
- GM organisms might pass their transferred genes on to other organisms — for example, weeds might gain the ability to resist glyphosate.
- GM plants and animals have been patented by companies that then restrict the availability of non-GM alternatives.

As a result of these public concerns, the use of recombinant DNA technology is regulated by law and, in the EU, is restricted.

> **Typical mistake**
>
> Students often give one-sided and heartfelt answers about the ethics of recombinant DNA technology. You must give a balanced account, demonstrating knowledge of A-level biology. The commonly seen expression 'playing at God' fails to show such knowledge or understanding.

Exam practice

1 The polymerase chain reaction (PCR) is used to amplify segments of DNA.
   (a) If one PCR cycle takes 5 minutes, tick the box that shows how many molecules of DNA will
   be produced from a single template molecule in 1 hour. [1]

   5
   12
   2^5
   2^{12}

   (b) The DNA polymerase used in the PCR must be able to withstand high temperatures.
      (i) Describe the role of the DNA polymerase used in the PCR. [?]
      (ii) Explain why a high temperature is used during the PCR. [2]
   (c) Explain why a primer is used in the PCR. [3]

2 A sample of DNA was sequenced using the dideoxynucleotide chain termination method.
   (a) Explain why labelled dideoxynucleotides are used in this sequencing method. [2]
   (b) The DNA fragments produced during this sequencing were separated by gel electrophoresis.
   The results are shown in the diagram.

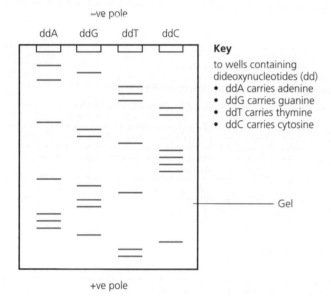

Key
to wells containing
dideoxynucleotides (dd)
• ddA carries adenine
• ddG carries guanine
• ddT carries thymine
• ddC carries cytosine

   (i) Explain why the fragments migrated from the top of the gel to the bottom of the gel shown
   in the diagram. [3]
   (ii) Use information in the diagram to deduce the sequence of the first six bases in the *template*
   DNA strand. Explain your answer. [2]
   (c) Give *two* ways in which scientists can use the results of gene sequencing. [2]

3 Compare and contrast the structures and roles of primers and probes in DNA technology. [4]

> **Exam tip**
>
> Question 4 tests content from another specification topic. Be
> prepared for synoptic questions like this in your A-level biology exam.

4 Read the passage below and then answer the questions that follow it.

Bacteria possess a natural defence mechanism against viruses. Part of their genome contains a base
sequence that repeats many times, referred to as clustered regularly interspaced short palindromic
repeats (**CRISPRs**). Within these repeats are spacers — unique sequences of about 20 bases that
match the DNA of viruses that have infected these bacteria.

Located near to the CRISPR sequences are genes that encode CRISPR-associated proteins (**Cas**). [5]
These Cas proteins are enzymes that cut DNA.

Bacteria transcribe the base sequence of each spacer into a molecule of RNA. Each Cas enzyme takes up one of the RNA molecules and cradles it. When such a Cas–RNA complex encounters matching DNA from a virus, the RNA binds to it and the Cas enzyme cuts both the viral DNA strand to which the RNA binds and the complementary strand of the DNA. [10]

Cas9 is an enzyme, produced naturally by the bacterium *Streptococcus pyogenes*, which is used in biotechnology. A Cas9–RNA complex can be used by researchers to produce knockout mice in a much shorter time than the traditional method.

(a) In the context of DNA, what are short palindromic repeats (lines 2–3)? [2]
(b) Use your knowledge of the lytic cycle of a virus to suggest an explanation for the origin of the spacers within the CRISPR parts of a bacterial genome (lines 3–4). [2]
(c) What name is given to a bacterial enzyme that cuts DNA (line 6)? [1]
(d) Contrast the action of Cas enzymes (lines 9–10) with the induced-fit model of enzyme action. [2]
(e) Suggest how scientists could use Cas9 enzymes to produce knockout mice (lines 12–13). [3]

Answers and quick quiz 7 online

ONLINE

Summary

Using gene sequencing

- The polymerase chain reaction (PCR) amplifies DNA samples, providing sufficient material for investigation, such as gene sequencing and DNA profiling.
- Gene sequencing involves finding the base sequence of an organism's genome.
- Once known, gene sequences can be used to predict the amino acid sequence of polypeptides and to find links with inherited conditions.

Factors affecting gene expression

- Transcription factors are proteins that are essential for the activation of genes.
- Transcription factors bind to the promoter region of a gene. The resulting DNA–transcription factor complex activates RNA polymerase, enabling the gene to be transcribed.
- Epigenetic modification produces changes in gene expression that can be passed from generation to generation. They do not, however, involve changes in the sequence of DNA bases. Instead, they affect the ability of RNA polymerase to transcribe a gene by methylation of cysteine-carrying nucleotides or by acetylation of the histones around which DNA is wound within chromosomes.

Stem cells

- Stem cells retain the ability to replicate themselves and give rise to other cell types. Epigenetic modifications are involved in a human totipotent stem cell (a zygote) becoming a pluripotent stem cell (a blastomere) and finally into a fully differentiated somatic cell.
- Stem cells offer the possibility in medicine of repairing or replacing damaged cells.
- The use of induced pluripotent stem cells (iPS cells) in medicine is less controversial than the use of embryonic stem cells.

Gene technology

- Recombinant DNA can be produced by combining DNA from two different organisms. Restriction endonucleases cut DNA from the two organisms and DNA ligase anneals the two DNA structures together.
- A variety of vectors can be used to transfer recombinant DNA into another organism, transforming the recipient, which becomes a genetically modified organism (GMO).
- Marker genes and replica plating are used in one technique for finding transformed bacteria.
- Recombinant DNA technology has been used to improve the qualities of a number of crop plants, including soya beans.
- The use of recombinant DNA technology and GMOs is controversial and tightly regulated.

8 Origins of genetic variation

This topic builds on content from Topic 2 (meiosis) and Topic 3 (natural selection).

Origins of genetic variation

Look back to Topic 2 to remind yourself of the following sources of genetic variation:

- Production of new alleles of genes by gene mutation (page 20).
- New combinations of alleles in gametes following:
  - random assortment of homologous chromosomes during anaphase I of meiosis (page 38)
  - crossing over between homologous chromosomes during prophase I of meiosis (page 38)
  - chromosome mutation (page 39)

Now test yourself

TESTED

1 Other than those listed, give one other source of genetic variation in sexually reproducing organisms.
2 Name the source(s) of genetic variation in bacteria.

Answers on p. 202

Transfer of genetic information

Terms used in genetics

REVISED

You must be able to define, and show understanding of, the following terms used in genetics.

Gene and allele

- A **gene** is a sequence of DNA nucleotide bases that encodes the amino acid sequence of a polypeptide.
- An **allele** is one of two or more different forms of a gene. Alleles have slightly different nucleotide sequences that will lead to the production of polypeptides with different properties.

Genotype and phenotype

- The **genotype** is the genetic constitution of an organism. Since diploid organisms have two copies of each chromosome, they will also have two copies of each gene, so we represent their genotype using two symbols, for example **AA**. Since diploid organisms produce their gametes by meiosis, the gametes only have one copy of each gene (i.e. are haploid). We represent the genotype of a gamete with only one symbol, for example **A**.
- The **phenotype** is the observable or measurable characteristic of an organism resulting from an interaction between the genotype and the organism's environment.

Homozygous and heterozygous

These terms refer to the nature of the two copies of each gene in an organism's genotype:

- In a **homozygous** genotype, both copies of a gene are the same allele, for example **AA** or **aa**. A diploid organism that has a homozygous genotype for a particular gene is called a **homozygote**.
- In a **heterozygous** genotype, the two copies of the same gene are different alleles, for example **Aa**. A diploid organism that has a heterozygous genotype for a particular gene is called a **heterozygote**.

Dominance, recessive and codominance

These terms refer to the way in which alleles exert their effect on the phenotype:

- A **dominant** allele always shows its effect in the phenotype, whether the organism is homozygous (**AA**) or heterozygous (**Aa**) for that gene. We usually represent a dominant allele of a gene with an upper case letter, for example **A**.
- The effect of a **recessive** allele is masked by the dominant allele in a heterozygote, so only shows its effect in the phenotype of a homozygous genotype (**aa**). We usually represent the recessive allele of a gene using a lower case letter, for example **a**.
- **Codominant** alleles are equally dominant; the influence of both alleles can be seen in the phenotype of a heterozygote.

Multiple alleles

- If a gene has more than two alleles, they are described a **multiple alleles**.

Inheritance of a sex-linked gene, using haemophilia as an example

REVISED

A-level biology courses cover the inheritance of two genes — **dihybrid inheritance** — and build on your study of genetics at GCSE, when you learned about the inheritance of only one gene (monohybrid inheritance).

The A-level course also covers the inheritance of sex-linked genes. Since you studied the inheritance of sex at GCSE, this is perhaps the simplest type of cross to remind yourself of genetic crosses and pedigree diagrams.

In mammals, sex is determined by sex chromosomes: a very short Y chromosome and a much longer X chromosome. As you saw on page 40:

- a female has two X chromosomes, represented as XX
- a male has one X chromosome and one Y chromosome, represented as XY

The long X chromosome has many genes that are absent from the short Y chromosome. Such a gene is said to be **sex-linked**.

Haemophilia is an inherited condition is which the time taken for blood to clot is much longer than usual. The condition is the result of a sex-linked gene with two alleles:

- a dominant allele (**H**) that results in fast blood clotting
- a recessive allele (**h**) that results in slow blood clotting

In mammals the gene is found on the X chromosome, so we can represent these alleles and the X chromosome together as X^H and X^h.

> A **sex-linked** gene is one that is present on only one of the sex chromosomes, X or Y.

Exam tip

An examiner might set a question involving sex-linkage in animals other than mammals. Do not assume that females are XX and males are XY in all animals; read the information in the question carefully.

Inheritance of haemophilia using a genetic cross

Figure 8.1 reminds you of the way you should represent a genetic cross. It shows a cross between a man and woman who both have fast blood clotting times. The woman, however, is heterozygous for the haemophilia allele (we refer to her as a 'carrier'). The genetic cross includes:

- clear titles of what each row represents
- a clear separation of the genotypes of individuals and those of gametes
- a Punnett square that reduces the risk of making an error when recording the possible genotypes of all the offspring that two parents might have
- a clear indication of the ratio of the phenotypes of offspring that might be expected of the parents

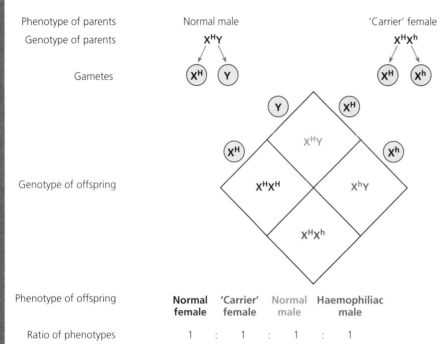

Exam tip

Even if you are confident about your knowledge of genetics, it is a good idea to use diagrams of genetic crosses to reduce the chance of making an error in an exam.

Exam tip

It is a good idea to get into the habit of encircling the genotypes of gametes. This makes the difference between gametes and individuals crystal clear to an examiner.

Figure 8.1 A genetic cross representing the inheritance of haemophilia — an example of sex linkage

Now test yourself

TESTED

3 What is the *probability* of the couple represented in Figure 8.1 having a haemophiliac son?
4 It is possible for a woman with fast blood clotting to have a haemophiliac daughter. Draw a labelled genetic diagram to explain how the daughter would inherit haemophilia.

Answers on p. 202

Inheritance of haemophilia using a pedigree diagram

Figure 8.2 shows an alternative way to represent inheritance — a pedigree diagram. Like the genetic cross in Figure 8.1, it has features that are standard:

- A female is represented by a circle and a male by a square.
- A horizontal line between a female and a male links two parents.
- A vertical line from the line linking two parents leads to their offspring.
- The offspring are shown in birth sequence from left to right.
- A code, in this case colour, is used to show the phenotype of each individual.

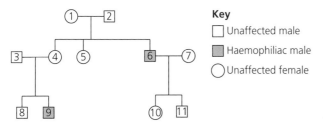

Figure 8.2 A pedigree diagram showing the inheritance of haemophilia in one family

Key
☐ Unaffected male
▣ Haemophiliac male
◯ Unaffected female

Exam tip

When using a pedigree diagram to work out the genotypes of individuals, always start with an individual showing the inherited disorder who has parents that do not show the inherited disorder.

TESTED ☐

Now test yourself

5 In Figure 8.2, individual 6 is male with the genotype X^hY.
  (a) Explain why this conclusion must be correct.
  (b) Give the genotypes of his parents, individuals 1 and 2.
  (c) Give the genotype of his sister, individual 4.

Answers on p. 202

Inheritance of two genes with autosomal linkage

REVISED ☐

Autosomal linkage occurs when two genes are located on the same **autosome**, as opposed to a sex chromosome.

An **autosome** is any chromosome other than a sex chromosome.

Now test yourself

TESTED ☐

6 What is the name given to the place on a chromosome where a particular gene is located.

Answer on p. 202

The fruit fly, *Drosophila melanogaster*, has four pairs of chromosomes; one pair of sex chromosomes and three pairs of autosomes. The following genes are both located on autosome 2:
● A gene controlling body colour, which has two alleles:
  ○ a dominant allele (**G**) that results in a grey body
  ○ a recessive allele (**g**) that results in a black body
● A gene controlling wing length, which has two alleles:
  ○ a dominant allele (**L**), which results in long wings
  ○ a recessive allele (**l**), which results in vestigial (very short) wings

Look back to Figure 2.7 (page 38) to remind yourself of the effect of crossing over during meiosis. Like humans, fruit flies produce gametes by meiosis. Figure 8.3 shows the effect of crossing over on the gametes that can be produced by a heterozygous fruit fly with the genotype **GgLl** in which the linked alleles are **GL** and **gl**.

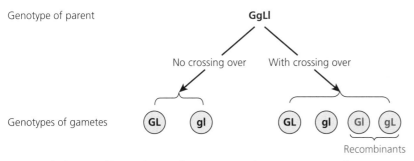

Genotype of parent **GgLl**

No crossing over   With crossing over

Genotypes of gametes   GL gl   GL gl Gl gL

Recombinants

Figure 8.3 The effect of crossing over on the genotypes of gametes

Inheritance of two linked genes in the absence of crossing over between them

Figure 8.4 represents a cross between two *D. melanogaster* that are homozygous for the body colour and wing length genes.
- The female (represented by ♀) is homozygous dominant for both genes, **GGLL**.
- The male (represented by ♂) is homozygous recessive for both genes, **ggll**.

Exam tip

It might help you to avoid errors when representing genotypes of gametes if you bracket the alleles of genes that are linked, for example (**GL**) and (**gl**).

Now test yourself

TESTED

7 Suggest why using the letter **C** to represent the gene for body colour and **W** to represent the gene for **w**ing length might lead to errors in an exam.

Answer on p. 202

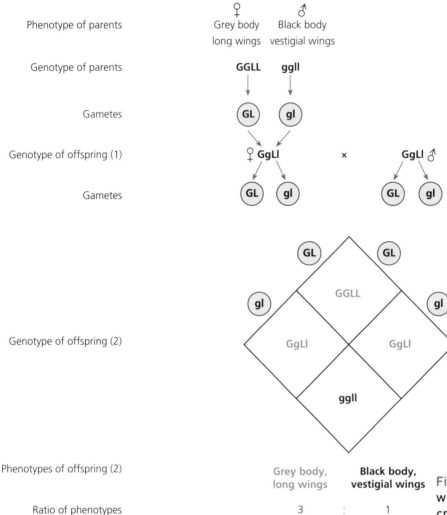

Figure 8.4 A genetic cross with autosomal linkage and no crossing over

You can see from Figure 8.4 that all the offspring of the two parent flies have the same genotype (GgLl) and so will have the same phenotype (grey body and long wings). One of the female offspring is then mated with one of the male offspring.
- In the absence of crossing over, both heterozygous parents produce gametes with the genotypes GL and gl.
- As a result of independent assortment and random fertilisation, we can expect equal numbers of each potential fertilisation to occur.
- Since two of the genotypes produce the same phenotype, we expect the offspring to show two genotypes in the ratio 3:1.

Exam tip

If all the offspring of a cross involving two linked genes have the same phenotypes as the parents, there has been no crossing over of the linked genes.

Now test yourself

TESTED

8 Other than the fact that the genes are linked and there has been no crossing over between them, give three assumptions that you would be making in expecting a ratio of 3:1 in Figure 8.4.

Answer on p. 203

Inheritance of two linked genes when crossing over occurs between them

Consider the *D. melanogaster* example of making a homozygous dominant female with a homozygous recessive male. If crossing over occurs between the gene for body colour and wing length, it will:

- occur only in a small number of meiotic divisions
- produce new (non-parental) combinations of alleles in some gametes, i.e. **Gl** and **gL**

Figure 8.5 represents a cross in which crossing over has occurred during gamete production in the female fly but has not occurred in the male fly.

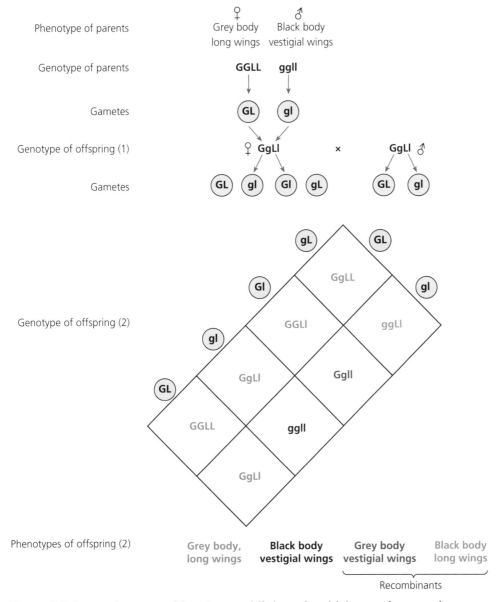

Figure 8.5 A genetic cross with autosomal linkage in which crossing over has occurred during gamete production in the female

As a result of crossing over, the 2nd offspring generation of Figure 8.5 includes individuals with genotypes that were not present in Figure 8.4. In turn, this leads to phenotypes that were not present in Figure 8.4 and, consequently, a ratio of phenotypes that is different from the expected 3:1 ratio.

Now test yourself

TESTED

9 Suggest the relationship between the frequency of crossing over between two genes and their distance apart on the chromosome.
10 You could use Figure 8.4 to calculate the expected ratio of phenotypes in the 2nd offspring generation. Explain why you could not do the same with Figure 8.5.

Answers on p. 203

> **Exam tip**
>
> If some offspring of a cross involving two linked genes have phenotypes, and hence genotypes, not seen in either parent, crossing over has occurred.

Inheritance of two non-interacting unlinked genes

REVISED

In pea plants, the characteristics flower colour and stem height are controlled by **non-interacting unlinked** genes.

The gene controlling flower colour has two alleles:
● A dominant allele (**F**) that results in violet flowers.
● A recessive allele (**f**) that results in white flowers.

The gene controlling stem height has two alleles:
● A dominant allele (**T**) that results in tall stems.
● A recessive allele (**t**) that results in short stems.

> Genes are said to be **non-interacting** if they control two unrelated phenotypic features.

> Genes are **unlinked** if their loci are located on different chromosomes.

Now test yourself

TESTED

11 Explain the term *unlinked* genes.

Answer on p. 203

Figure 8.6 shows a genetic cross involving homozygous parent plants, one with violet flowers and tall stems and the other with white flowers and short stems.

The offspring from the first cross are allowed to interbreed to produce a 2nd offspring generation. As a result of independent assortment and random fertilisation:
● The 1st offspring generation are all heterozygous (**FfTt**) with the phenotype violet flowers and tall stems.
● The 2nd offspring generation show all four possible combinations of flower colour and stem height in the ratio 9:3:3:1.

> **Exam tip**
>
> If you find a ratio of 9:3:3:1 in the 2nd generation of a cross in an exam question you should immediately recognise you are dealing with two non-interacting unlinked genes.

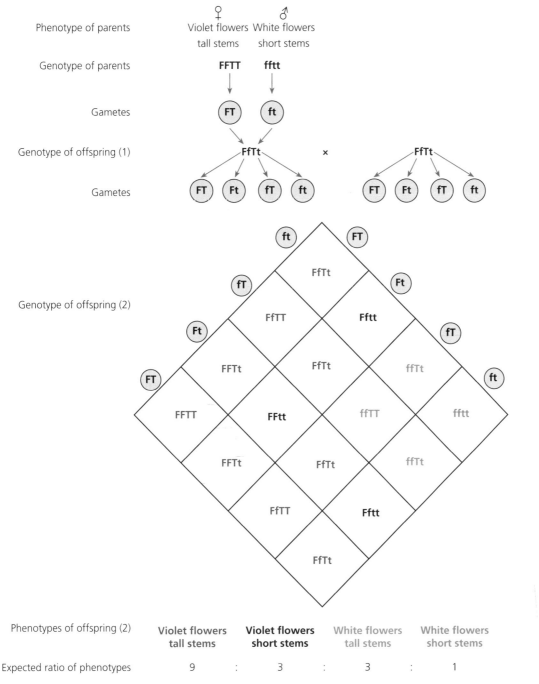

Figure 8.6 **A genetic cross between homozygous pea plants involving two non-interacting unlinked genes**

Using the chi squared (χ^2) test of significance

REVISED

The results of genetics crosses seldom match exactly the predictions you expect from an understanding of the pattern of inheritance. How much discrepancy between the results you expect and the results you obtain can you tolerate before accepting that the model on which you based your expectations is wrong? Look at Table 8.1, which contains data from a cross between pea plants.

● The row labelled 'Observed' shows the number of the four different phenotypes that a biologist obtained from a cross like that in Figure 8.6.
● The row labelled 'Expected' shows the number of each phenotype the biologist expected from her understanding of this pattern of inheritance.

Now test yourself TESTED

12 Explain how the biologist obtained the 'expected' values in Table 8.1.

Answer on p. 203

Table 8.1 The results obtained from a cross involving pea plants and the results expected from Figure 8.6

Results	Frequency of each phenotype				Total
	Violet flowers and tall stems	Violet flowers and short stems	White flowers and tall stems	White flowers and short stems	
Observed	173	66	55	26	320
Expected	180	60	60	20	320

The data in Table 8.1 are **categoric**, so you test whether the difference between the expected results and observed results is purely due to chance using the chi squared (χ^2) test. To do this you use the following formula:

$$\chi^2 = \sum \frac{(O - E)^2}{E}$$

where $\sum$ means 'sum of', O represents the observed results and E represents the expected results.

Table 8.2 expands the data in Table 8.1 to show how the data are used to calculate the value of χ^2.

> **Categoric** data are those where only certain values can exist, in this case the frequency of each phenotype.

Table 8.2 Calculating the value of χ^2

Values	Phenotype			
	Violet flowers and tall stems	Violet flowers and short stems	White flowers and tall stems	White flowers and short stems
O	173	66	55	26
E	180	60	60	20
$(O - E)$	–7	6	–5	6
$(O - E)^2$	49	36	25	36
$\dfrac{(O - E)^2}{E}$	0.27	0.60	0.42	1.80
χ^2	3.09			

To find what this value means, you need to check it against a table of critical values for χ^2. Part of such a table is shown in Table 8.3.

Table 8.3 Critical values for χ^2

Degrees of freedom	Value of χ^2 at a probability value (p) of		
	0.05	0.01	0.001
3	7.81	11.34	16.27

Table 8.3 requires some explanation.

> **Typical mistake**
>
> Students often write that a statistical test examines whether their results are due to chance. It is the *difference* between their results and what they predicted that might be due to chance.

- You are testing whether the difference between your results and the results you expected from an understanding of the theory is due to chance alone.
- There are four categories of data (the four phenotypes) in Table 8.2. Since you know that the total was 320 plants, once you have values for any three categories there can only be one value for the fourth category. This is referred to as '**degrees of freedom**' and, in this case, there are three degrees of freedom.
- The probability values are decimal values of 1.0 (certainty). By common consent, scientists agree on a probability value at which they conclude that the differences between what they expected and what they got was not due to chance. Usually that cut-off point is a probability value equal to or less than 0.05 ($p \leq 0.05$). This represents a chance of 1 in 20.

Looking at the values in Table 8.3, you can see that the value of χ^2 calculated in Table 8.2 is much less than the value at a **probability level** of 0.05 with three degrees of freedom. From this you can conclude that it is highly likely that the difference between your results and what you expected was due to chance. Another way of putting this is that the difference between the observed and expected results is not **significant**.

Degrees of freedom reflect the freedom of values to remain unknown. If we know the total value, or mean value, of several observations, say x observations, as soon as we know the value of all but one of the observations, we know the value of the final observation, so the data have $(x - 1)$ degrees of freedom.

A **probability level** is a probability value that we accept represents an event that is so unlikely that we can reject the notion that it is caused by chance alone.

A difference between two sets of data is **significant** if the probability of it occurring by chance alone is equal to, or less than, the probability level we have agreed to use.

Now test yourself

TESTED ☐

13 Explain why there are three degrees of freedom in Table 8.3.

Answer on p. 203

Gene pools

Natural selection can change allele frequencies in a population

REVISED ☐

You learned about natural selection in Topic 3 (page 57). Differential reproductive success changes the frequency of alleles of genes within the **gene pool** of a population. As a result of natural selection:
- the frequency of a beneficial allele of a gene increases
- the frequency of a disadvantageous allele of a gene decreases

Figures 8.7 and 8.8 show the effects of stabilising selection and disruptive selection on populations.

> A **gene pool** is all the alleles of all the genes present in a particular population at a given time.

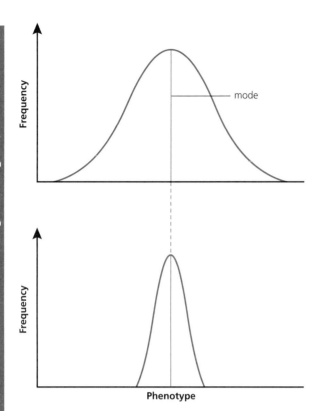

Figure 8.7 Stabilising selection maintains continuity by reducing variation around a modal value

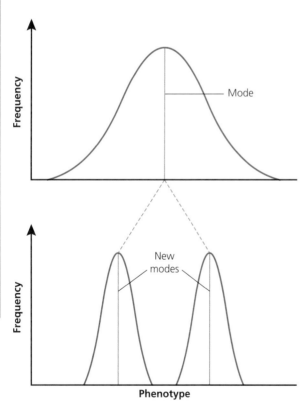

Figure 8.8 Disruptive selection leads to two or more phenotypes within a population (called polymorphism), which might result in speciation

TESTED

Now test yourself

14 Define the term 'differential reproductive success'.
15 Use your knowledge from Topic 3 (e.g. pages 57 and 58) to explain how disruptive selection might lead to speciation.

Answers on p. 203

Changes in allele frequencies can be the result of chance events

REVISED

Whether you land a 'head' or a 'tail' when you toss a coin in the air is a chance event. You are more likely to get a 50:50 split of 'heads' and 'tails' if you toss a coin 1000 times than if you toss the same coin 10 times.

TESTED

Now test yourself

16 Explain why you are more likely to get a 50:50 split of 'heads' and 'tails' if you toss a coin 1000 times than if you toss it 10 times.

Answer on p. 203

The same is true of changes in allele frequencies in populations. Chance is less likely to affect the allele frequencies in a gene pool if the population is large than if it is small. The process by which changes in allele frequency occur by chance in small populations is called **genetic drift**.

> **Genetic drift** refers to changes in allele frequency that occur by chance in small populations.

Two events that lead to genetic drift are population bottlenecks and the founder effect. A **population bottleneck** occurs when a population is drastically reduced in size by some environmental factor, such as disease.
- The variety of alleles in the gene pool of the survivors is much less than that of the original population.
- If the population recovers, its gene pool is that of the few survivors.

Figure 8.9 models a population bottleneck using different shapes to represent four genes and different colours to represent the alleles of these genes.

TESTED

Now test yourself

17 Cheetahs have a similar immune system to humans. Suggest why a skin graft from one cheetah to another is unlikely to be rejected by the recipient's immune system.

Answer on p. 203

In the **founder effect**, a few individuals colonise a new environment and breed there.
- Only a fraction of the alleles of genes present in the population is present in these few colonising individuals.
- As these individuals reproduce, the gene pool of the new population will be that of the colonisers and not that of the population from which they came.

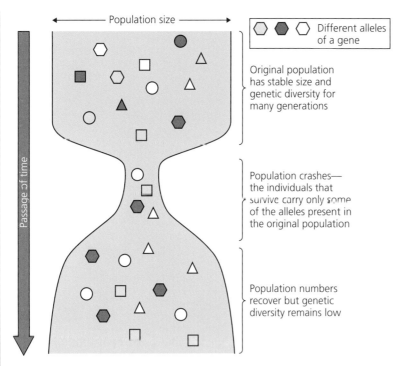

Figure 8.9 During a population bottleneck, few alleles of each gene are represented among the survivors of the population crash

Now test yourself

TESTED

18 Populations of mice on the Scottish islands show much less genetic diversity than those on the mainland of Scotland. Suggest what caused this low genetic diversity.

Answer on p. 203

The Hardy-Weinberg equation

REVISED

The **Hardy-Weinberg equation** allows you to:
● use observations of phenotype frequencies to calculate the associated **allele frequencies** in a population
● monitor whether allele frequencies are changing in a population

> The **Hardy-Weinberg equation** is $p^2 + 2pq + q^2 = 1$, where p is the frequency in the gene pool of the dominant allele and q is the frequency of the recessive allele.
>
> **Allele frequencies** measure how common alleles of a gene are in a population. The frequencies of the alleles of a single gene are expressed as decimal values.

Now test yourself

TESTED

19 The Hardy-Weinberg equation is based on the Hardy-Weinberg principle that allele frequencies do not change from generation to generation. Suggest *two* assumptions that are essential for this principle to be true.

Answer on p. 203

Typical mistake

Students often confuse the Hardy-Weinberg principle and the Hardy-Weinberg equation. The principle tells us that, under certain conditions, allele frequencies should not change from generation to generation. The latter enables us to calculate allele frequencies in a population.

Now test yourself

TESTED ☐

20 Within a population, the frequencies of the alleles of a single gene must add up to 1.0. Explain why.

Answer on p. 203

If the gene has two alleles, say **A** and **a**:
- The unknown frequency of allele **A** in the gene pool is represented by p.
- The unknown frequency of allele **a** in the gene pool is represented by q.
- Their frequencies must add up to 1.0, i.e. $p + q = 1$.
- there are three possible genotypes in the population, **AA**, **Aa** and **aa**. This leads us to the Hardy-Weinberg equation, which tells us that:

$$p^2 + 2pq + q^2 = 1$$

Now test yourself

TESTED ☐

21 In a population, 16% of the organisms showed the phenotype of a homozygous recessive genotype.
  (a) Represent 16% as a decimal frequency value.
  (b) Use your answer to part (a) to give the frequency of the recessive allele and of the dominant allele.
  (c) What is the frequency of the heterozygotes in this population?
22 How could you use the Hardy-Weinberg equation to tell whether a population is evolving?

Answers on p. 203

> **Exam tip**
>
> When using the Hardy-Weinberg equation, always start with the frequency of the homozygous recessive individuals. They are the only ones you can recognise from their phenotype; you know that they have the genotype **aa** and that their frequency is q^2.

Exam practice

1 (a) Distinguish between the terms genotype and phenotype. [2]
   Hair colour and hair length are two features of a dog's fur.
   The gene for hair colour has an allele for dark colour and an allele for albino colour. The gene for hair length has an allele for long hair and an allele for short hair.
   A breeder repeatedly crossed a male and female, both of which had dark, short hair. The features of the fur of their offspring are shown in the table.

Feature of fur	Number of offspring
Albino, long hair	2
Albino, short hair	8
Dark, long hair	10
Dark, short hair	29

  (b) What can you conclude about the inheritance of these two genes from the results shown in the table? Use a genetic diagram to justify your answer. [5]

2 The pedigree shows the inheritance of red-green colour blindness over four generations of one family.

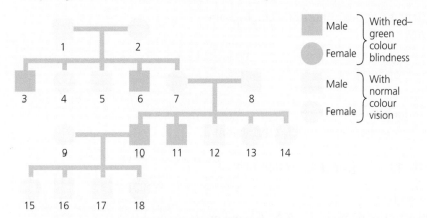

(a) Explain the evidence in the pedigree that shows:
  (i) the gene for colour vision is sex linked [1]
  (ii) the allele for red-green colour blindness is recessive [1]
(b) Give the genotype of individual 7. Explain your answer. [3]
(c) Can you conclude from the pedigree that individual 9 has a different genotype from her mother-in-law, individual 7? Explain your answer. [3]

3 The gene encoding the enzyme phenylalanine hydroxylase has two alleles. The recessive allele results in an inability to produce the enzyme, a condition called phenylketonuria (PKU). Carriers of this allele are symptomless.
(a) Explain why carriers of the allele for PKU are symptomless. [2]
(b) In the UK, 9 in 10 000 people have this condition. What is the frequency of carriers in the UK? Explain your answer. [3]

4 Rice paddies are level, flooded fields used to grow rice. In many countries, rice paddies are weeded by hand when the rice plants are young.
Early watergrass, *Echinochloa oryzoides*, is a common weed in paddy fields. It is referred to as a rice mimic because it is unlike other species of *Echinochloa* but closely resembles rice plants.
(a) Suggest how this rice mimic might have evolved. [4]
E. oryzoides was first found in paddy fields in New South Wales, Australia, in 1938. A gene sequencing analysis of ten rice-field populations of *E. oryzoides* from paddy fields in New South Wales showed that they were genetically uniform for the 32 loci tested.
(b) Suggest a reason for the genetically uniform populations of *E. oryzoides* in New South Wales. [2]

Answers and quick quiz 8 online

ONLINE

Summary

Genetic variation

- Genetic variation refers to the number of alleles of each gene present in the gene pool of a population.
- Mutations are the source of new genetic variation and the sole source of genetic variation in populations of organisms that reproduce asexually.
- In sexually reproducing populations, further genetic variation results from random assortment of homologous chromosomes and crossing over during meiosis, and from random fertilisation of gametes.

Transfer of genetic information

- During sexual reproduction, adult organisms produce haploid gametes that fuse to produce a diploid zygote.
- Diagrams to represent genetic crosses show the genotypes of parents, gametes and offspring in a way that explains the origin of the offspring phenotypes. Pedigree charts represent only the phenotypes, from which the genotypes must be deduced.

- The alleles of unlinked genes are inherited separately; those of linked genes are inherited together unless crossing over has occurred during gamete formation.
- The ratio of phenotypes in the offspring of a cross can indicate whether two genes are linked or unlinked.
- Sex linkage results from a gene being located on a sex chromosome.

Gene pools

- Natural selection results in changes in the frequency of the alleles of a gene within a population.
- Stabilising selection reduces the genetic variation within a gene pool; disruptive selection causes a polymorphism within a population that might lead to speciation.
- In small populations, chance events can cause changes in the allele frequencies within a gene pool. This is called genetic drift and includes population bottlenecks and the founder effect.
- The Hardy-Weinberg equation ($p^2 + 2pq + q^2 = 1$) can be used to monitor changes in the allele frequencies within a population.

9 Control systems

This topic builds on content from Topic 4 (circulation and transport).

Homeostasis

In biology, **homeostasis** refers to the ability of organisms to keep conditions within their bodies relatively stable. They do this by maintaining a dynamic equilibrium, i.e. corrective mechanisms ensure that despite large fluctuations in the external environment only small variations occur in the internal environment.

The three homeostatic mechanisms you studied maintain a relatively stable:
- blood pH — changes in pH affect enzyme activity
- core body temperature — changes in temperature affect enzyme activity
- blood water potential — changes in water potential can cause cell lysis, cell shrinkage and disruption to hydrolytic metabolic reactions

> **Homeostasis** is the maintenance of a state of dynamic equilibrium within the body of an organism.

> **Typical mistake**
>
> Students often define homeostasis in terms of maintaining constant conditions in the body. The internal conditions are not constant, but they are relatively stable.

Now test yourself

TESTED

1 Suggest the importance of maintaining a relatively constant concentration of glucose in your blood.

Answer on p. 203

Positive and negative feedback

REVISED

Homeostatic mechanisms involve positive and negative feedback loops.

Negative feedback systems maintain relative stability

Negative feedback systems reverse any departure from an ideal condition in the internal environment. The key steps in negative feedback are as follows:
- A change occurs in the internal environment.
- This change is detected by receptors.
- The receptors lead to activation of a mechanism that reverses the change.
- The conditions return to 'ideal' and the corrective mechanism is switched off.

Figure 9.1 shows that changes above and below the 'ideal' value can be corrected. Since the mechanisms that correct increases and decreases from the 'ideal' value are constantly switched on and off, negative feedback systems maintain a state of **dynamic equilibrium**.

> **Negative feedback** is a process that ensures that any departure from an ideal state results in a return to the ideal state.

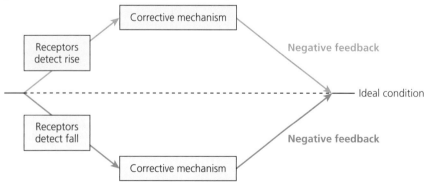

Figure 9.1 Negative feedback corrects changes in either direction from an ideal condition

Exam practice answers and quick quizzes at **www.hoddereducation.co.uk/myrevisionnotes**

Positive feedback systems produce change

The key steps in **positive feedback** are as follows:
- A change occurs in the internal environment.
- This change is detected by receptors.
- The receptors lead to activation of a mechanism that continues the change.

Positive feedback can result when negative feedback systems fail, for example, when suffering hypothermia. They can also occur in a beneficial way, for example the release of cytokines by activated T helper cells.

> **Positive feedback** causes a departure from a starting condition to lead to a further departure from that condition.

Now test yourself

TESTED

2 Explain why the release of cytokines by activated T helper cells is an example of positive feedback.

Answer on p. 203

Chemical control in mammals

Chemical control in mammals can involve:
- **cytokines** — chemicals released by cells in the immune system that stimulate other cells in the immune system (page 121)
- **neurotransmitters** — chemicals released at the ends of neurones that stimulate their target cells (page 164)
- **hormones** — chemicals released by ductless endocrine glands and released into the bloodstream, through which they reach their target cells.

In each case, the target cells have **receptor proteins** on their cell surface membranes that are complementary to the chemical to which they respond.

Mode of action of hormones

REVISED

Although hormones are carried by the blood to all body tissues, they only affect those with receptor proteins on their cell surface membranes that are complementary to the hormone molecule.

Hormones affect their target cells in one of two ways.

Peptide hormones and the second messenger mode of hormone action

Hormones that are peptide molecules do not enter their target cells. Instead they act as 'first messengers', stimulating 'second messengers' that are inside the target cells.

Figure 9.2 shows this **second messenger model** of action, using the hormone glucagon as an example.
- A molecule of the hormone attaches to a receptor protein on the cell surface membrane of its target cell.
- This binding activates an enzyme on the inside of the cell surface membrane (the G-protein in Figure 9.2 is one such enzyme).
- The activated enzyme hydrolyses ATP, removing two of its phosphate groups to form a molecule of cyclic adenosine monophosphate (cAMP).
- The cAMP activates further enzymes that cause a cascade of reactions in the cell. In the example in Figure 9.2, the result is the formation of glycogen from glucose.

> In the **second messenger model** of action, the binding of a hormone to a specific receptor protein on the outside of its target cell activates a 'second messenger' inside the cell.

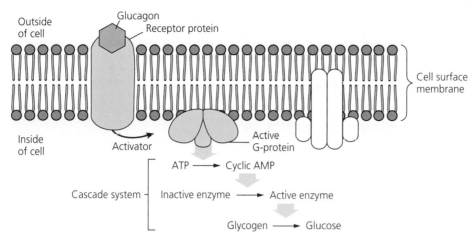

Figure 9.2 **The action of glucagon illustrates the second messenger model of hormone action**

The peptide hormone adrenaline acts in a similar way but binds to different receptor molecules and activates different enzymes within the target cell.

Now test yourself

TESTED

3 Name the type of reaction from which glycogen is made from glucose.
4 Name the type of cell that is the target for the hormone glucagon.

Answers on p. 203

Steroid hormones enter cells and bind to transcription factors

Hormones that are steroid molecules are able to cross the cell surface membranes of their target cells. Once inside the cells, they bind to transcription factors.

Figure 9.3 represents the action of these steroid hormones. The numbered stages are described on the next page.

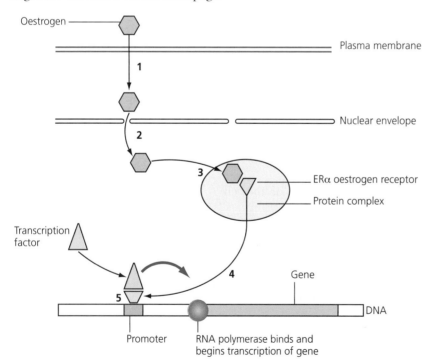

Figure 9.3 **Steroid hormones, such as oestrogen, enter their target cells and stimulate the transcription of target genes**

1 Oestrogen diffuses through the cell surface membrane of its target cell.
2 Once inside the cell, oestrogen diffuses into the cell's nucleus.
3 In the nucleus, oestrogen binds to an oestrogen receptor (called ERα) that is contained within a protein complex.
4 This causes the oestrogen receptor to change shape and leave the protein complex that was inhibiting its action.
5 The activated oestrogen receptor binds to the promoter region of a gene and attracts other transcription factors to bind to it. As a result, RNA polymerase begins to transcribe the target gene.

Now test yourself

TESTED

5 Steroids are a type of lipid. How does this enable steroid hormones to cross the cell surface membranes of their target cells?
6 What is a transcription factor?

Answers on p. 203

Chemical control in plants

Plant growth substances

REVISED

Chemical control of many plant responses is brought about by **plant growth substances**. You need to be familiar with the effects of the three plant growth substances shown in Table 9.1. Each of these is produced by cells in the **meristems** of roots and shoots.

> **Meristems** are regions of actively dividing cells such as those found at the tips of roots and shoots.

Table 9.1 **The major effects of three classes of plant growth substance**

| Process | Effect of named plant growth substance | | |
	Auxins	Cytokinins	Gibberellins
Growth of stem	At relatively high concentrations, promote cell elongation	Promote cell division	If auxins are also present, promote cell elongation
Growth of root	At relatively high concentrations, inhibit cell elongation	No effect	No effect
Growth of lateral roots	Promote root formation	No effect	No effect
Suppression of growth of lateral buds (apical dominance)	Suppress growth of lateral buds	Promote growth of lateral buds	Reinforce effect of auxins in suppressing growth of lateral buds

Now test yourself

TESTED

7 Plant growth substances often interact with one another (are agonistic). Use Table 9.1 to find:
   (a) a pair of plant growth substances that have agonistic actions
   (b) a pair of plant growth substances that have antagonistic actions
8 Horticulturalists often dip stem cuttings into rooting powder to encourage root growth. Use Table 9.1 to suggest which class of plant growth substance is present in rooting powder.

Answers on p. 203

Phytochrome

REVISED

Phytochrome is not a plant growth substance. It is a blue–green photoreceptor pigment that occurs in two forms (Figure 9.4):
- P_R — the inactive form in which phytochrome is produced
- P_{FR} — the active form that results when P_R absorbs red light

Sunlight contains more red light than far-red light, so the longer the day (or, conversely, the shorter the night) the more P_{FR} will accumulate in plant tissues.
- P_{FR} is the active form of phytochrome.
- It is transported from the cytoplasm into the nucleus of plant cells.
- Here it activates transcription factors that, in turn, stimulate the transcription of genes affecting **photomorphogenesis**.

Phytochrome and flowering

One important aspect of photomorphogenesis is the change from vegetative growth to flowering. Species of plants respond in one of two ways to P_{FR}:
- Flowering in **long-day plants** is promoted by P_{FR}. These plants will only flower if the night is short or if a long night is interrupted by a brief period of light.
- Flowering of **short-day plants** is inhibited by P_{FR}. These plants will only flower if the night is long and uninterrupted by a period of light.

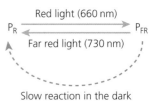

Figure 9.4 **Phytochrome occurs in two forms**

> **Photomorphogenesis** describes the effect of light on the growth and development of plants

Now test yourself

TESTED

9 Suggest why long-day plants might be more accurately called 'short-night plants'.

Answer on p. 203

Structure and function of the mammalian nervous system

The nervous system is composed of nerves

REVISED

Nerve cells, or **neurones**, are adapted to:
- transmit impulses along their length
- secrete chemicals, called **neurotransmitters**, directly onto their target cells

Three main types of neurone occur:
- **Sensory neurones** transmit impulses from receptors to relay neurones.
- **Connector**, or relay, **neurones** transmit impulses within the central nervous system.
- **Motor neurones** transmit impulses to effectors, such as muscles and glands.

Large numbers of neurones are located within **nerves**:
- Sensory nerves contain only sensory neurones.
- Motor nerves contain only motor neurones.
- Mixed nerves contain both sensory and motor neurones.

> **Neurones** are cells that are specialised to transmit nerve impulses.
>
> **Neurotransmitters** are substances secreted by one end of a neurone that cross a synapse and affect depolarisation of another neurone.
>
> **Nerves** carry the axons of a vast number of neurones, which are supported by connective tissue.

Typical mistake

Students often confuse the terms 'nerve' and 'nerve cell'. It helps to avoid this confusion if you get into the habit of using the term 'neurone' when referring to the cell.

The nervous system is composed of a central and a peripheral nervous system

REVISED

Figure 9.5 shows how the components of your nervous system are classified.

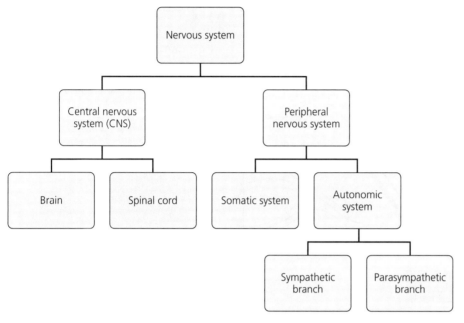

Figure 9.5 The major components of the mammalian nervous system

Your **central nervous system** comprises your:
● **brain**, located within your skull
● **spinal cord**, running from your brain to the base of your backbone

Localisation of function within your brain

Figure 9.6 shows four of the major areas within your brain. One of the main functions of each is described below.

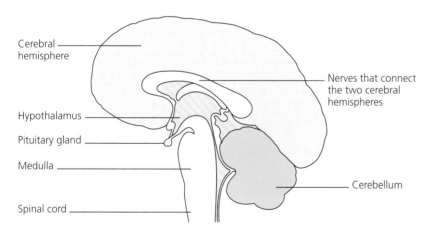

Figure 9.6 The location of four major areas of the mammalian brain
● **Cerebellum** — controls involuntary movements, such as balance and coordination.
● **Cerebrum** — consists of two cerebral hemispheres and is the site of decision making, for example, initiating movement.
● **Hypothalamus** — the control centre of the autonomic nervous system, controlling temperature regulation and osmoregulation.
● **Medulla oblongata** — controls breathing and heart rate.

Structure of your spinal cord

Figure 9.7 shows the location of two areas of tissue within your spinal cord.

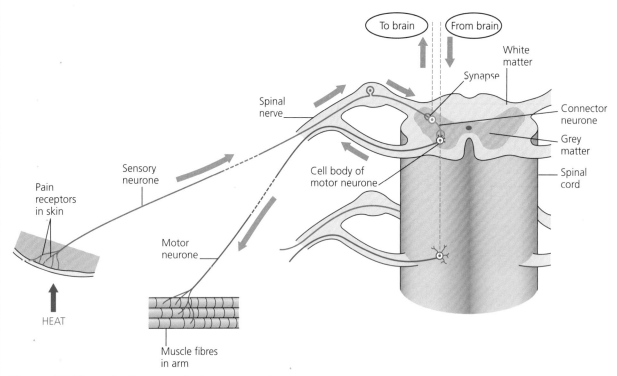

Figure 9.7 The spinal cord, spinal nerves and the location of neurones within them

- **White matter** — contains the myelinated axons of motor neurones leaving and sensory neurones entering the cord.
  - ○ Sensory neurones enter via a branch from a spinal nerve that is at the back (dorsal side) of the cord, called the **dorsal root**.
  - ○ Motor neurones leave via a branch from a spinal nerve that is at the front (ventral side) of the cord, called the **ventral root**.
- **Grey matter** — contains unmyelinated neurones that:
  - ○ link sensory and motor neurones in the **reflex arcs** that control unlearnt, reflex actions
  - ○ run up and down the spinal cord to and from the brain

Now test yourself

TESTED

10 Suggest what causes the white matter of the spinal cord to appear white.

Answer on p. 203

Somatic and autonomic parts of your peripheral nervous system

Somatic system

The **somatic system** controls conscious responses to external stimuli.
These include:
- responses to sensations, such as touch or pain
- voluntary movement of your skeletal muscles

Exam practice answers and quick quizzes at **www.hoddereducation.co.uk/myrevisionnotes**

Autonomic system

The **autonomic system** controls responses to internal stimuli over which you have no conscious awareness or control. These include:
- balance
- heart rate

The autonomic system has two branches that act antagonistically:
- Neurones of the **sympathetic branch** release the neurotransmitter noradrenaline. This branch prepares your body for action in times of stress — for example, the 'fight or flight' responses.
- Neurones of the **parasympathetic branch** release the neurotransmitter acetylcholine. This branch produces the opposite effect from the sympathetic branch, calming and conserving energy.

Nervous transmission

Neurones

REVISED

Figure 9.8 shows how a motor neurone is adapted for its function.
- The **cell body** contains the nucleus and most of the cell organelles.
- Extensions from the cell body, called **dendrons** (which have smaller extensions called **dendrites**) are stimulated by neurotransmitters released by an adjacent sensory or relay neurone (Figure 9.7).
- The **axon** is a long extension from the cell body. It transmits impulses from the cell body to a cell within a target organ — a muscle or a gland.
- The **myelin sheath** around the axon insulates the axon, restricting movement of ions across its cell surface membrane to the small gaps between cells forming the sheath.
  - **Myelin** is a lipid contained in the cell surface membrane of **Schwann cells**, which wrap themselves around the axons as they develop.
  - The small gaps between Schwann cells are called **nodes of Ranvier**.
- The **synaptic knobs** release neurotransmitter onto cells of the target organ, stimulating their response.

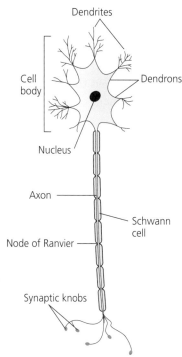

Figure 9.8 **A motor neurone**

Now test yourself

TESTED

11 What is the difference in function of the dendron and the axon of a motor neurone?

Answer on p. 203

Axons have a resting potential

REVISED

When an axon is not transmitting an impulse, it is **polarised**, meaning that there is a difference in charge between its cytoplasm and the external medium. This results in a **membrane potential** of about $-70\,mV$. This is called the **resting potential**, and is maintained in the following way:
- Active transport of sodium ions (Na^+) out of and potassium ions (K^+) into the cell occurs through carrier proteins in the cell surface membrane, called **sodium–potassium pumps**. For every three Na^+ pumped out only two K^+ are pumped in; the cytoplasm becomes more negative as it loses cations.

> The **membrane potential** of a cell is the potential difference (voltage) between the cytoplasm and the external medium.

- Facilitated diffusion of K^+ out of the cell and of Na^+ into the cell occurs through open K^+ channels and Na^+ channels in the surface membrane. The Na^+ diffuse in more slowly than the K^+ diffuse out.
- Large, negatively charged particles (mainly proteins) remain in the cytoplasm.

Exam tip

Examiners will award a mark if you write that axons transmit impulses. They will not if you refer to axons 'carrying messages' or 'carrying information'.

Typical mistake

Students often explain polarisation and depolarisation in terms of the movement of sodium and potassium. It is sodium ions and potassium ions that are involved, and you must say so.

Development of an action potential

REVISED

When a receptor or another neurone secretes neurotransmitter across an excitatory **synapse** it causes the membrane potential of the receiving cell to change momentarily. The cytoplasm becomes positive with respect to the outside, with a potential of about +40mV. This is called the **action potential**.

Voltage-gated ion channels and the action potential

In addition to the ion channels already mentioned, the cell surface membranes of neurones have **voltage-gated ion channels**. Some of these are voltage-gated Na^+ channels and some are voltage-gated K^+ channels.

Figure 9.9 shows the involvement of these voltage-gated ion channels in the development of, and recovery from, an action potential.

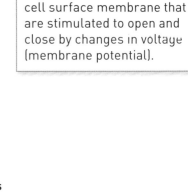

A **synapse** is the tiny gap between two neurones.

Voltage-gated ion channels are protein channels in a cell surface membrane that are stimulated to open and close by changes in voltage (membrane potential).

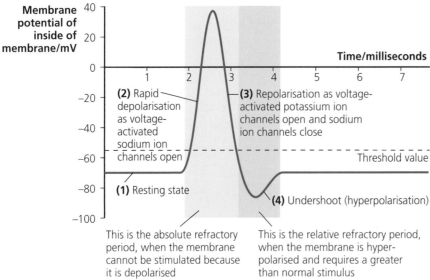

Figure 9.9 Depolarisation and repolarisation at a point on the axon produce a single action potential

The numbers on the graph refer to the following events.

1 Initially, all the voltage-gated Na^+ channels and voltage-gated K^+ channels are closed.
2 Some voltage-gated Na^+ channels open, allowing sodium ions to diffuse into the axon. This makes its membrane potential less negative. As the membrane potential reaches and then exceeds a **threshold level**, of about −55mV, more and more voltage-gated Na^+ channels open. Sodium ions rapidly diffuse into the axon. As a result, its membrane potential rises to about +40mV. This is the 'spike' you can see in Figure 9.9. Reaching the threshold membrane potential results in the **all-or-nothing** response of neurones.

Neurones operate on an **all-or-nothing** principle. Either an action potential occurs or it does not. If it does, an action potential is always the same no matter how big the stimulus.

3 A membrane potential of +40 mV stimulates two events:
   ○ The voltage-gated Na⁺ channels close — the membrane potential cannot become any more positive.
   ○ The voltage-gated K⁺ channels open — potassium ions rapidly diffuse out of the axon.
   This restores the resting potential of about −70 mV.
4 In practice, too many potassium ions diffuse out of the axon, reducing the membrane to about −80 mV. This 'undershoot' is called **hyperpolarisation**.

> **Typical mistake**
>
> Students tend to write about ions 'rushing' into or out of an axon. It is better use of terminology to write about 'rapid diffusion'.

Now test yourself

TESTED

12 Explain how the threshold membrane potential of −55 mV results in the all-or-nothing response of neurones.
13 Is the all-or-nothing response an example of negative feedback or of positive feedback? Explain your answer.

Answers on p. 203

The refractory period

After an action potential has started at a particular point on an axon membrane, there is a period of time when it is impossible for another depolarisation to occur at that same point. This is called the refractory period and is also shown in Figure 9.9.

● The **absolute refractory period** (shown in yellow) includes the periods of depolarisation and repolarisation.
● The **relative refractory period** (shown in pink) includes the period of hyperpolarisation and the time during which the 'resting' balance of sodium and potassium ions is restored.

Propagation of an impulse along an axon

REVISED

An impulse moves along an axon because an action potential at one point on the axon causes depolarisation of the adjacent downstream membrane (Figure 9.10).

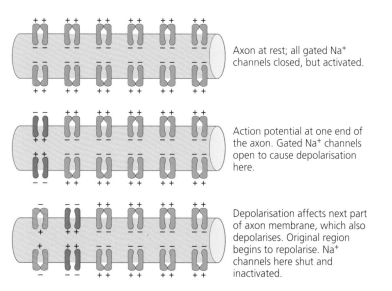

Axon at rest; all gated Na⁺ channels closed, but activated.

Action potential at one end of the axon. Gated Na⁺ channels open to cause depolarisation here.

Depolarisation affects next part of axon membrane, which also depolarises. Original region begins to repolarise. Na⁺ channels here shut and inactivated.

⊓ Gated Na⁺ channel closed, but activated   ⊓⊓ Gated Na⁺ channel open   ⊓ Gated Na⁺ channel closed and inactivated

Figure 9.10 Impulse propagation in an unmyelinated axon

- At the point on the axon where the action potential occurs, the inside of the axon is positive with respect to the outside.
- As a result, negatively charged ions ahead of the point of the action potential move back to this positive area.
- This causes an action potential in the membrane just ahead of the first action potential.
- The action potential cannot go backwards because the membrane immediately behind the action potential is in its own refractory period.

> **Exam tip**
>
> You must ensure that you can explain how an action potential is formed and how it is propagated along an axon. Write a bullet list of the main stages and revise them regularly.

Now test yourself

TESTED

14 What causes negatively charged ions to move backwards to the point of an action potential in Figure 9.10?
15 Use information in Figures 9.9 and 9.10 to explain why an impulse can only travel in one direction.

Answers on p. 203

Myelinated axons and saltatory conduction

Ions can only cross the membrane of a myelinated axon where it is exposed at the nodes of Ranvier. As a result, the depolarisations 'jump' from node to node. This is called **saltatory conduction** after the Latin word *saltare*, meaning 'to jump'.

Saltatory conduction enables myelinated axons to transmit impulses much faster than unmyelinated axons.

> **Saltatory conduction** refers to the rapid conduction of action potentials in myelinated axons as they 'jump' from node of Ranvier to node of Ranvier.

Synapses

REVISED

A synapse is the junction between two neurones — a **pre-synaptic neurone** and a **post-synaptic neurone**.

Structure of a synapse

The components of a synapse are shown in Figure 9.11.
- The **synaptic knob** of the pre-synaptic cell has:
  ○ a surface membrane (the **pre-synaptic membrane**) with gated calcium ion channels
  ○ cytoplasm containing a large number of mitochondria
  ○ vesicles containing a neurotransmitter; in Figure 9.11 this is acetylcholine
- The **synaptic cleft** is a gap of about 20 nm between the two cells.
- The dendrite of the post-synaptic neurone has a cell surface membrane (the **post-synaptic membrane**) with:
  ○ neurotransmitter receptors that act as gated sodium ion channels
  ○ gated potassium ion channels

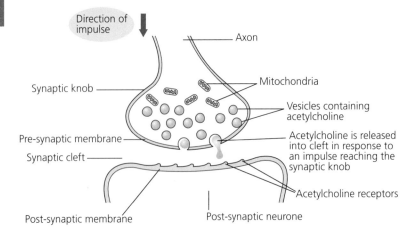

Figure 9.11 A cholinergic synapse uses acetylcholine as its neurotransmitter

Transmission of an impulse across a synapse

Figure 9.12 shows the sequence of events that occurs as an impulse is transmitted across:

- an **excitatory synapse**, which stimulates an impulse in the post-synaptic cell
- an **inhibitory synapse**, which inhibits an impulse in the post-synaptic cell

These synapses have different effects because the pre-synaptic cells release different neurotransmitters. For example, the pre-synaptic cells in many excitatory synapses release acetylcholine whilst those in inhibitory synapses release noradrenaline.

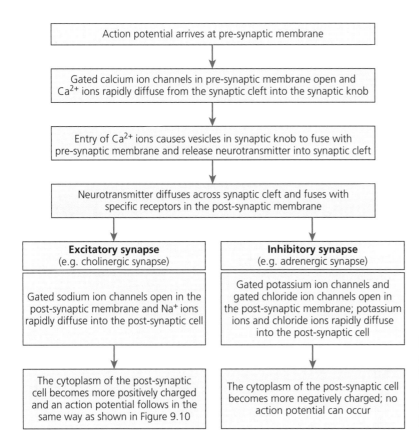

Figure 9.12 The events at an excitatory synapse and at an inhibitory synapse. In cholinergic synapses the neurotransmitter is acetylcholine; in adrenergic synapses it is noradrenaline

Stopping synaptic transmission

If the pre-synaptic cell continued to secrete its neurotransmitter, transmission across the synapse would be long-lived.

This does not happen because, in addition to receptors for neurotransmitter, the post-synaptic membrane contains molecules of an enzyme that hydrolyses the neurotransmitter. In cholinergic synapses, this hydrolytic enzyme is called **acetylcholinesterase**.

- Acetylcholinesterase hydrolyses acetylcholine into acetate and choline:

$$\text{acetylcholine} \xrightarrow{\textit{acetylcholinesterase}} \text{acetate} + \text{choline}$$

- The choline is taken up by the pre-synaptic neurone and is used to form acetylcholine again:

$$\text{acetylcoenzyme A} + \text{choline} \xrightarrow{\textit{choline acetyltransferase}} \text{acetylcholine} + \text{coenzyme A}$$

- At the same time, calcium ions are removed from the synaptic knob back into the synaptic cleft by active transport.

As a result, the pre-synaptic cell reverts to its resting position again and synaptic transmission stops.

Effects of drugs on the nervous system

Many drugs and poisons exert their effects in the body by influencing synaptic transmission. You are expected to be able to recall the following three examples.

- **Cobra venom**, a deadly poison, blocks the acetylcholine receptors in the post-synaptic membranes of cholinergic synapses.
- **Lidocaine**, an anaesthetic often used for local pain relief, blocks the voltage-gated sodium ion channels in the post-synaptic membranes of synapses.
- **Nicotine**, the addictive component of tobacco, mimics the effect of acetylcholine in cholinergic synapses.

Detection of light by mammals

Structure of the mammalian retina

REVISED

Photoreceptor cells in the retina of the eyes of mammals contain pigments that degrade when struck by light. This degradation enables us to see. Figure 9.13 shows the position of these photoreceptors.

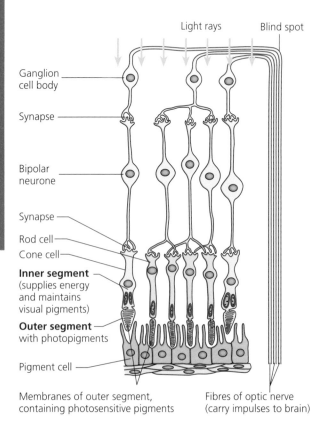

Figure 9.13 The location of rods and cones in the mammalian retina

Table 9.2 summarises the key features of the two types of photoreceptor shown in Figure 9.13.

Table 9.2 A comparison of cone cells and rod cells

Feature	Cone cells	Rod cells
Distribution within retina	Concentrated at centre (the **fovea**) with few at periphery	Concentrated at periphery with none at the fovea
Photosensitive pigment	Iodopsin	Rhodopsin
Sensitivity of pigment to light	Iodopsin degraded only by light of relatively high intensity	Rhodopsin degraded even by light of low intensity
Sensitivity to colour	Three types of cone cell, each sensitive to red light or to green light or to blue light (**trichromatic theory** of colour vision)	Insensitive to colour
Nature of synapses to bipolar cells	Synapse individually with bipolar cells that synapse individually with ganglion cells (a type of sensory neurone), resulting in high **visual acuity**	Synapse in groups with individual bipolar cells that also synapse in groups with a single sensory neurone (**retinal convergence**) resulting in low visual acuity

Now test yourself

TESTED ☐

16 You learned about resolving power in Topic 2. What does it mean in the context of visual acuity?

Answer on p. 203

> **Visual acuity** refers to the resolving power of rods and cones.

Role of rhodopsin in initiating action potentials in ganglion cells

REVISED ☐

The rod cells in Figure 9.13 are involved in a pathway involving three types of cell (rod cell, bipolar cell, ganglion cell) and two types of synapse (inhibitory synapse and excitatory synapse).

When light strikes the photosensitive pigments of rod cells, the pigment is degraded. This results in an action potential in a ganglion cell within the optic nerve.

When *not* being stimulated by light:
- The cell surface membranes of the rod cells are depolarised.
- As a result, rod cells release their neurotransmitter onto the bipolar cell with which they synapse.
- As this is an inhibitory synapse, the bipolar cell does not release its neurotransmitter onto a ganglion cell.
- No impulses are sent from the ganglion cell through the optic nerve to the brain.

When stimulated by light:
- The pigment within rods is degraded into opsin and retinal.
- Opsin stimulates a series of enzyme-catalysed reactions that cause the cell surface membrane of the rod cell to become hyperpolarised.

- The rod cells no longer release their neurotransmitter onto the bipolar cell.
- The cell surface membrane of the bipolar cell becomes depolarised and releases its neurotransmitter onto a ganglion cell.
- The ganglion cell becomes depolarised and transits an impulse through the optic nerve to the brain.

In the dark, rhodopsin is resynthesised:

$$\text{opsin + retinal} \xrightarrow{\quad\text{ATP}\quad\text{ADP}\quad} \text{rhodopsin}$$

Now test yourself   TESTED ☐

17 On a dark night, you see a star out of the corner of your eye. When you look directly at it, it seems to have disappeared. Explain why.

Answer on p. 203

Control of heart rate in mammals

This builds on what you learned about the heart and the cardiac cycle in Topic 4.

You will have noticed that your heart rate increases when you are exercising and slows when you are resting. Your heart is responding to two changes in your blood:

- Its carbon dioxide concentration, detected by **chemoreceptors** in your aorta and your carotid arteries.
- Your blood pressure, detected by **baroreceptors**, also in your aorta and your carotid arteries.

Impulses from these receptors pass to a **cardiovascular control centre** in your medulla oblongata. This cardiovascular centre is divided into two antagonistic centres: the **acceleratory centre** and the **inhibitory centre**.

The medulla, in turn, sends impulses to the sinoatrial node (SAN) of your heart. Figure 9.14 summarises the effect on the SAN of changes in the carbon dioxide concentration of the blood.

Now test yourself   TESTED ☐

18 Give the precise location of:
   (a) the medulla oblongata
   (b) the SAN

Answers on p. 204

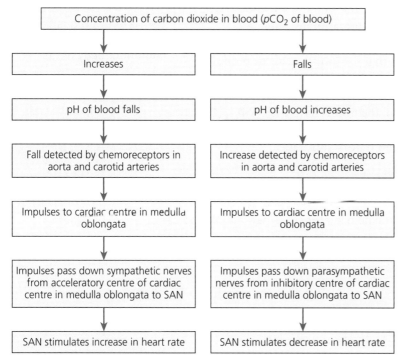

Figure 9.14 The effect on the heart rate of changes in carbon dioxide concentration of the blood

Revision activity

Using Figure 9.14 as a template construct a flow diagram to summarise the effects of increases and decreases in blood pressure.

Osmoregulation and temperature regulation

Osmoregulation in mammals

REVISED

Maintaining the water balance of cells is important if they are to metabolise effectively. Mammals lose water in their exhaled breath, sweat, faeces and urine. They replace it with water that comes from drinking and eating water-based food, and from metabolic reactions. Much of the water loss cannot be controlled (e.g. exhaled air), and could lead to a change in the water content of their body fluids.

Mammals regulate the water content of their body fluids by the effects of an interplay between:
- the **hypothalamus**, which contains osmoreceptors and produces antidiuretic hormone (**ADH**)
- the **posterior pituitary gland**, which stores and secretes ADH produced by the hypothalamus
- water reabsorption in the kidneys

Now test yourself

TESTED

19 What type of metabolic reaction is a source of water for mammals?

Answer on p. 204

In addition to their role in **osmoregulation**, the kidneys are also the major organs of nitrogenous excretion. In mammals, the main nitrogenous excretory product is **urea**.

- Urea is produced by cells in the liver that use amino acids to produce glucose, a process called **gluconeogenesis**.
- The amino group (NH_2) is removed from each amino acid (**deamination**) and is converted within the liver cells to urea, $(NH_2)_2CO$.
- The remaining part of the deaminated amino acid (CH_2RCOOH) is used to make glucose and used as a respiratory substrate.

> **Osmoregulation** is the regulation of the water content of body fluids.

Now test yourself

TESTED

20 Amino acids are normally used for protein synthesis. When might it be beneficial to use them in gluconeogenesis instead?

Answer on p. 204

Structure of the mammalian kidney

Mammals have two kidneys located in the lower abdomen. Figure 9.15(a) shows the gross structure of a kidney. It has a number of distinct regions — the **cortex**, **medulla** and **pelvis** — and contains vast numbers of microscopic tubules, called **nephrons**.

> **Nephrons** are the microscopic tubules that are the functioning units of mammalian kidneys.

(a)

LS through kidney showing positions of nephrons in cortex and medulla

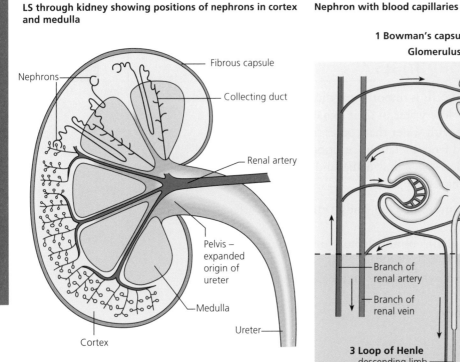

(b)

Nephron with blood capillaries

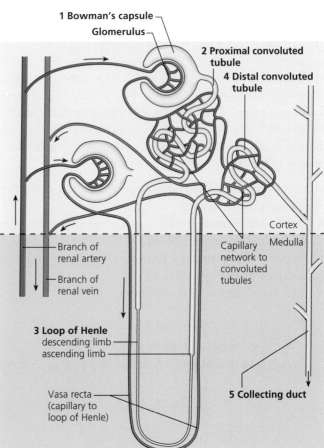

Roles of the parts of the nephron:

1 Bowman's capsule + glomerulus = ultrafiltration
2 proximal convoluted tubule = selective reabsorption from filtrate
3 loop of Henle = water conservation
4 distal convoluted tubule = pH adjustment and ion reabsorption
5 collecting duct = water reabsorption

Figure 9.15 (a) The gross structure of a mammalian kidney (b) A single nephron

Functioning of the mammalian kidney

Each nephron filters small metabolites from the blood, reabsorbs those that are useful and passes the remainder to collecting ducts that eventually empty into the bladder as urine. Use Figure 9.15(b) to follow the numbered descriptions of nephron function below.

1 Ultrafiltration in the renal capsule

Each **Bowman's capsule** is served by a capillary network, called a **glomerulus**. You learned about the formation of tissue fluid in Topic 4 (pages 84–85). In much the same way as tissue fluid is formed, water, ions and small molecules are forced out of the capillaries of the glomerulus into the lumen of the Bowman's capsule. As with tissue fluid formation, the filtration membrane is the **basement membrane** on which the cells lining each capillary sit.

The high blood pressure that forces these particles out of the capillaries is caused by contraction of the ventricular walls and enhanced by the arteriole leaving each glomerulus having a much narrower diameter than the arteriole leading into it.

The fluid forced from each glomerulus passes through gaps between the cells lining the capsule (called **podocytes**) into the lumen of the Bowman's capsule. This fluid is the **ultrafiltrate**.

Now test yourself

TESTED

21 Use your knowledge of tissue fluid formation to suggest the blood components that will be present in the ultrafiltrate in a Bowman's capsule.
22 Podocytes, the specialist cells lining the Bowman's capsule, have gaps between them. Suggest the advantage of these gaps.

Answers on p. 204

2 Selective reabsorption in the proximal convoluted tubule

The first part of each nephron is the **proximal convoluted tubule**. It reabsorbs from the ultrafiltrate:
- some water, by osmosis
- all the glucose and amino acids, by co-transport with sodium ions
- many ions, by facilitated diffusion and active transport
- some urea, by diffusion

These substances are taken up by the dense capillary network around each proximal convoluted tubule.

3 The loop of Henle acts as a countercurrent multiplier

The loops of Henle lie mainly in the medulla of each kidney. Their activity results in the medulla having a very negative water potential.
- Cells lining the upper part of each ascending limb actively secrete sodium ions and chloride ions from the ultrafiltrate into the tissues of the medulla.
- Cells lining the upper part of each descending limb, opposite and close to this part of the ascending limb, are permeable to these ions. Sodium ions and chloride ions diffuse into this part of the descending limb.
- The processes above greatly increase the concentration of sodium and chloride ions in the tissues of the medulla. This is why they are referred to as a **countercurrent multiplier**.

A **countercurrent multiplier** increases the concentration of a substance flowing within tubes in opposite directions

- The high concentration of sodium ions and chloride ions results in the tissues of the medulla having a very negative water potential.
- As a result, water passes by osmosis from the collecting ducts (which have walls permeable to water) into the medulla and into the vasa recta capillaries around the loops of Henle.
- The result is ultrafiltrate that is more concentrated than the blood and hence urine that is more concentrated than the blood.

4 Selective reabsorption in the distal convoluted tubule

The distal convoluted tubule absorbs:
- some ions, by facilitated diffusion and active transport
- some water, by osmosis

The permeability of the cells lining this part of the tubule is affected by hormones (unlike those lining the proximal convoluted tubule). This allows fine control of reabsorption and is important in, for example, adjusting the pH of the body fluids.

5 Water reabsorption in the collecting duct

Water reabsorption by the collecting duct is the main process by which the water content of the blood (and so of the whole body) is regulated. Figure 9.16 summarises the homeostatic mechanism by which the hypothalamus, posterior pituitary gland and ADH regulate the water content of the blood.

> **Exam tip**
>
> Don't forget the importance of the capillary networks around the nephrons.

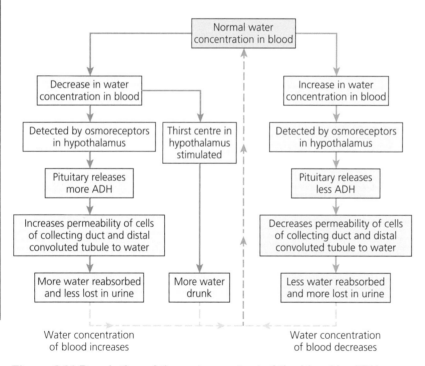

Figure 9.16 Regulation of the water content of the blood by ADH

Temperature regulation

You learned in Topic 1 that the rate of enzyme-catalysed reactions is affected by temperature changes. Consequently, animals that can regulate their body temperatures will be at an advantage over those that cannot.

Ectotherms and endotherms

Animals can be classed into two groups according to how they regulate their body temperatures:

- **Ectotherms** rely entirely on behavioural methods to gain or lose heat.
- **Endotherms** use physiological mechanisms as well as behavioural methods to gain or lose heat.

The mechanisms by which endotherms gain heat are:

- respiration — any energy released during respiration that is not used to produce ATP is released as heat
- conduction — heat gained by contact with objects warmer than the body
- convection — heat gained from air that is warmer than the body
- radiation — heat radiated from a warmer object

The mechanisms by which endotherms lose heat are:

- conduction — heat lost by contact with a cooler object
- convection — heat lost to cooler air
- radiation — heat radiated to a cooler environment
- evaporation — heat lost from the skin surface in converting sweat into vapour

> **Typical mistake**
>
> Sweating does not have a cooling effect. It is the evaporation of sweat that removes heat from the skin.

Thermoregulation in endotherms

The **core body temperature** of an endotherm is regulated by a **thermoregulatory centre** in the **hypothalamus**. It is divided into two antagonistic centres: the **heat gain centre** and the **heat loss centre**.

> The **core body temperature** is the temperature inside the trunk of an endotherm's body. Whilst skin temperature might change considerably, endotherms maintain their core body temperature.

The thermoregulatory centre receives impulses from **thermoreceptors** in two locations:

- in the skin, where they detect changes in skin temperature
- in the hypothalamus itself, where they detect changes in the blood temperature

> **Exam tip**
>
> Be sure to write that thermoreceptors detect *changes* in blood temperature. Do not write that they measure temperature; they are not thermometers.

The hypothalamus regulates body temperature via the autonomic nervous system.

- The heat gain centre stimulates actions that reduce heat loss and increase heat generation.
- The heat loss centre stimulates actions that increase heat loss and reduce heat generation.

Figure 9.17 summarises how it does this.

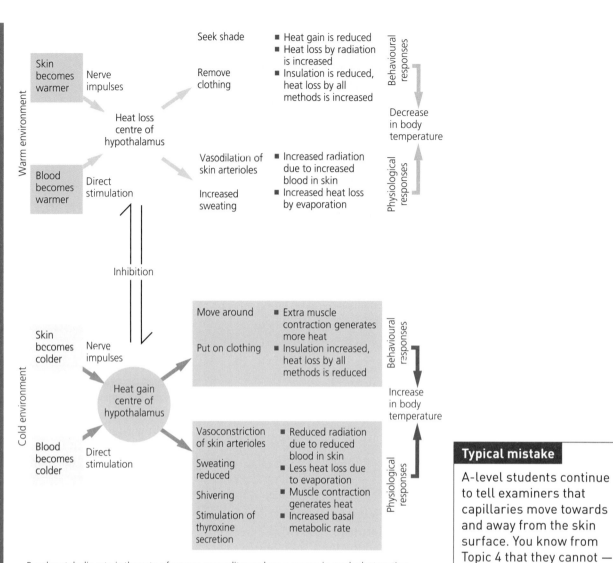

Warm environment

Cold environment

Basal metabolic rate is the rate of energy expenditure when a person is awake but resting, has not eaten for 12 hours and is comfortably warm.

Typical mistake

A-level students continue to tell examiners that capillaries move towards and away from the skin surface. You know from Topic 4 that they cannot — they have no muscles.

Figure 9.17 **Regulation of core body temperature in humans**

Exam practice

Exam tip

To answer question 1, you must use your understanding of other topics from the specification.

1 A kangaroo rat is a small animal that is well adapted to living in the hot, dry conditions of deserts. The table shows the water balance of a kangaroo rat over a period of 1 day.

Method of water gain	Volume/cm³ day⁻¹	Method of water loss	Volume/cm³ day⁻¹
Dry food	6.0	Evaporation from lungs	43.9
Metabolic water	54.0	Faeces	2.6
		Urine	13.5
Total	60.0		60.0

(a) Describe *two* ways in which a kangaroo rat can gain metabolic water. [4]

(b) A kangaroo rat spends most of the day underground, where the air is cooler and more humid than it is at the surface. Use information from the table to explain how this behaviour is an adaptation to living in a hot, dry desert. [2]

(c) The nephrons of a kangaroo rat have very long loops of Henle. Explain how this is an adaptation to living in a hot, dry desert. [4]

2 A student investigated photoperiodism in one species of plant. His results are shown in the diagram below.

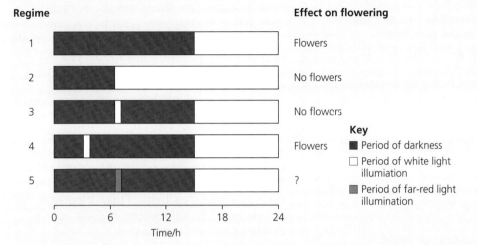

Regime   **Effect on flowering**

1   Flowers

2   No flowers

3   No flowers

4   Flowers

5   ?

Key
- ■ Period of darkness
- □ Period of white light illumiation
- ■ Period of far-red light illumination

(a) What can you conclude from the results of regimes 1 and 2 about the photoperiod of this species of plant? Justify your answer. [2]

(b) Explain the results from regimes 3 and 4. [3]

(c) What result would you expect from regime 5? Explain your answer. [3]

3 Many animals produce poisons. The table shows the effect of the poisons produced by two animals.

Poison produced by animal	Effect on synapses
Tetrodotoxin, produced by puffer fish	Blocks voltage-gated sodium ion channels
ω-conotoxin, produced by cone snails	Prevents calcium ions crossing pre-synaptic membranes

(a) Use your understanding of impulse transmission to explain why tetrodotoxin causes paralysis. [4]

(b) Use your understanding of cholinergic synapses to describe the effect of ω-conotoxin. [4]

(c) Conotoxins are short polypeptides. Some have been used as painkillers.
   (i) Suggest why conotoxins might be effective as painkillers. [1]
   (ii) When used as painkillers, conotoxins cannot be taken by mouth but must be injected. Suggest why. [1]

4 Many homeostatic mechanisms in mammals involve negative feedback control by hormones.
   (a) Explain the meaning of the term *negative feedback*. [2]
   (b) Compare the modes of action by which oestrogen and adrenaline affect their target cells. [4]
   (c) Oestrogen-dependent breast cancer is associated with over-production of oestrogen, stimulating cells to divide rapidly. A drug called tamoxifen is a successful treatment for this type of cancer. The molecular structure of this drug is similar to that of oestrogen.
   Use this information to explain how tamoxifen reduces the growth of oestrogen-dependent cancers. [2]

Answers and quick quiz 9 online

ONLINE

Summary

Homeostasis

- Homeostasis is the maintenance of a dynamic equilibrium of critical conditions in the body. These conditions include the core body temperature and the pH and water potential of the blood.
- Negative feedback corrects any deviation from the 'ideal' internal body state and is important in homeostatic mechanisms.

Chemical control in mammals and in plants

- Mammalian endocrine organs secrete hormones into the blood. They attach to specific receptor proteins on the target cells.
- Hormones affect their target cells either by binding to transcription factors or by initiating a second messenger mechanism within the cell.
- Plant growth substances include auxins, cytokinins and gibberellins. They are all produced by meristematic tissues.
- Auxins affect cell elongation and growth of lateral roots and buds. Auxins and cytokinins have an antagonistic effect on the growth of lateral buds.
- Phytochrome affects photomorphogenesis, including flowering. The active form of phytochrome is P_{FR}, which, in dim light, is slowly converted to the inactive form P_R.

The mammalian nervous system and nervous transmission

- The mammalian nervous system consists of the central and peripheral nervous systems. The latter is further divided into the somatic and autonomic nervous systems (ANS). The sympathetic branch of the ANS prepares the body for action whilst the parasympathetic branch has the opposite effect.
- Localisation of function occurs in the cerebrum, cerebellum, hypothalamus and medulla oblongata.
- Neurones are the functional unit of the nervous system. An impulse is carried by a neurone when it propagates an action potential along its length. Each action potential results from a change in membrane potential caused by the opening of voltage-gated ion channels.
- Myelinated neurones carry impulses faster than unmyelinated neurones by restricting to the nodes of Ranvier the points at which action potentials can occur.

- Synapses — the junctions between neurones — are bridged by the secretion of neurotransmitters that cross the synaptic cleft and affect the post-synaptic neurone. Excitatory synapses stimulate action potentials in post-synaptic membranes, whereas inhibitory synapses prevent action potentials in post-synaptic neurones. Cobra venom, lidocaine and nicotine exert their effect on synapses.
- Light is detected by rod cells and cones cells in the mammalian retina. The distribution of these cells within the retina, the nature of their synapses with neurones leading to the cerebrum and sensitivity of their pigments to light result in differences in visual acuity, colour vision and night vision.

Control of heart rate, water potential of blood and body temperature

- The cardiac centre of the medulla oblongata controls the heart rate of mammals via sympathetic and parasympathetic stimulation of the SAN in response to changes in the pressure and carbon dioxide concentration of their blood.
- Release of adrenaline by the autonomic nervous system also increases the heart rate.
- Mammals regulate the water potential of their blood by controlling the absorption of water by the collecting ducts in their kidneys. Homeostatic regulation is by the hypothalamus, which is able to detect the water potential of the blood and respond by controlling secretion of the hormone antidiuretic hormone (ADH) from the posterior pituitary gland.
- In controlling its core body temperature, an ectotherm relies on behavioural mechanisms to gain heat from its environment. In contrast, an endotherm is also able to gain heat from its own metabolic processes and can use behavioural and physiological mechanisms to control its core body temperature.
- The hypothalamus of an endotherm, such as a mammal, is able to detect changes in the temperature of its blood supply and regulate the core body temperature via the autonomic nervous system. A mammal is also able to detect temperature changes in its skin.

10 Ecosystems

The nature of ecosystems

What is meant by the term ecosystem?

REVISED

The term **ecosystem** describes the biological **community** living in one **habitat** at a particular time *and* the interactions of this community with its environment. These interactions might involve:

- **abiotic factors** — the chemical and physical features of the environment, such as availability of water, calcium concentration in the soil or light intensity
- **biotic factors** — interactions between organisms of different **populations** (interspecific) or organisms within one population (intraspecific)

Ecosystems can be quite small, for example a small pond, or vast, for example a tropical rainforest.

Now test yourself

TESTED

1 Distinguish between the following terms:
   (a) population and community
   (b) ecosystem and habitat

Answers on p. 204

> An **ecosystem** refers to the community living in a habitat and its abiotic and biotic interactions.
>
> A **community** describes all the populations living in one habitat at a particular time.
>
> The **habitat** is one part of an ecosystem in which a community lives, for example a clearing within a forest ecosystem is a habitat.
>
> A **population** is all the organisms of one species living in the same habitat at a particular time.

> **Exam tip**
>
> Examiners will expect you to use the terms community, ecosystem, habitat and population in an ecological sense and not as they would be used in everyday language.

Trophic levels within ecosystems

REVISED

An ecosystem is organised into **trophic levels**, including:

- **autotrophs**, usually referred to as **producers** because they use either photosynthesis or chemosynthesis to produce organic molecules from inorganic sources
- **heterotrophs**, usually referred to as **consumers** because they feed on other organisms

> **Tropic levels** are feeding levels within a community.

Transfer of biomass between trophic levels

There is a perpetual transfer of **biomass** within an ecosystem. This is represented in Figure 10.1.

- When a producer is eaten by a primary consumer, biomass is transferred from the producer level of the ecosystem to the consumer level of the ecosystem.
- When a primary consumer is eaten by a secondary consumer, biomass is transferred from the primary consumer level of the ecosystem to the secondary consumer level.

> **Biomass** is the mass of the inorganic and organic matter in an organism or within a trophic level. This is also often referred to as **dry mass**, i.e. excluding the mass of water.

- This transfer continues through the other trophic levels of the ecosystem, as shown in Figure 10.1.
- Ultimately, all organisms, or their waste products, pass to the decomposer trophic level, which hydrolyses them all to inorganic molecules or ions. This results in cycles of the inorganic components of biological molecules, such as the carbon cycle and nitrogen cycle you learnt during your GCSE course.

Now test yourself

TESTED ☐

2 What is the advantage of using dry mass rather than wet mass as a measure of biomass?

Answer on p. 204

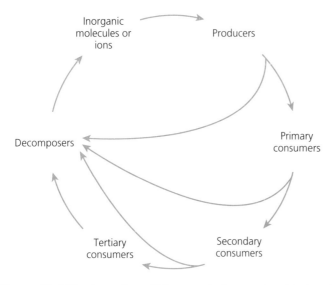

Figure 10.1 The transfer of biomass between trophic levels results in cycles of the inorganic components of biomass

> **Typical mistake**
>
> Students often report that nutrients are recycled as elements, for example carbon and phosphorus. They are not; they are recycled as inorganic compounds (e.g. carbon dioxide) or as ions (e.g. phosphates).

Now test yourself

TESTED ☐

3 To which trophic level do herbivores belong?
4 Name the *two* biological classification groups to which decomposers belong.

Answers on p. 204

Investigating ecosystems

REVISED ☐

When investigating natural habitats, ecologists aim to estimate:
- the **abundance** of organisms
- the **distribution** of organisms

The methods they use will depend on the nature of the organisms in the population(s) they are studying.

> **Abundance** is an indication of the number, or frequency, of organisms in a habitat.
>
> **Distribution** is the precise location of organisms within their habitat.

Methods for assessing the abundance of sessile organisms

Sessile organisms do not move, either ever (e.g. plants) or at the time of sampling (e.g. limpets exposed on a rocky shore). There are three commonly used methods for estimating the abundance of sessile organisms in a population:

- **Individual counts**, in which every individual of that population is counted. This method is suitable when the organisms are reasonably large and their numbers are low (e.g. limpets on a rocky shore).
- **Percentage cover**, in which an estimate is made of the area covered by members of a population. This method is suitable when the organisms cannot be counted individually, for example:
  - they form a more or less continuous cover on the substratum (e.g. lichens on a stone wall)
  - they are very small and their numbers are very large (e.g. barnacles on a rocky shore)
- **ACFOR** scales, which give a subjective estimate of the frequency of organisms in a given area. ACFOR is an acronym for the frequency groups shown in Table 10.1.

Table 10.1 Examples of how ACFOR scales might be used on two populations

Frequency group	Percentage cover of lichens on a stone wall/%	Number of periwinkles on a rocky shore/m^{-2}
Abundant	> 80	> 200
Common	50–79	100–199
Frequent	20–49	50–99
Occasional	10–19	10–49
Rare	< 10	< 10

Now test yourself

TESTED

5 Which measure of abundance would you use to investigate:
   (a) lichens on a tree trunk?
   (b) dandelion plants in a lawn?

Answers on p. 204

Sampling methods

It is normally impossible to measure the abundance of entire populations within a habitat. Instead, ecologists take samples that they hope represent the true size of the population, so that they can make inferences from them.

For a sample to be representative of the population from which it was taken it must be free of any bias by the investigators — it must be a **random sample**.

To sample a natural habitat, ecologists need:
- a device that determines the size of each sample
- a method to randomise the samples

> A **random sample** is a sample that is taken from a population without bias, i.e. without a conscious choice by the person taking the sample.

Frames used in sampling

Two types of frame are commonly used in sampling populations:
- **Quadrat frames** are squares, commonly 1 m × 1 m, but they can be 0.5 m × 0.5 m or even less. Once placed at random, you count the number of organisms, or estimate their percentage cover, within the frame. To make estimations of percentage cover easier, the frames are often subdivided into smaller squares by lengths of string, as in Figure 10.2.
- **Point frames** are bridge-like structures made from three pieces of wood with ten metal pins inserted through the horizontal arch of the bridge (Figure 10.3). After placing the point frame in the randomly chosen sampling spot, you lower each of the ten metal pins and record what it touches.

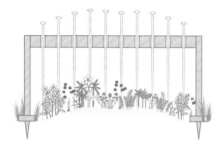

The frame is randomly placed a large number of times, the 10 pins lowered in turn onto vegetation and the species (or bare ground) recorded.

Figure 10.3 A point frame in use on a sand dune

Randomising samples using a sampling grid

You use this method in an area where you suspect that environmental variables do not change over the sampling area, for example in a field. The procedure is then as follows:

1 Devise a grid over a map of the area to be sampled and give the grid coordinates (Figure 10.4).
2 Replicate the grid on the area to be sampled, for example using two tape measures at right angles.
3 Use a random number generator on a computer or programmable calculator to produce a pair of coordinates, for example A5, B6.
4 Place a quadrat with its top left-hand corner at the intersection of the coordinates in the sample area.
5 Count the organisms or estimate the percentage cover of organisms within the quadrat.
6 Repeat using further pairs of coordinates until you have sufficient samples to be confident that you have a true representation of the population.
7 Calculate the mean number or percentage cover of organisms per unit area.

Typical mistake

From their exam answers, students seem to love the idea of throwing quadrat frames over their shoulders. This is not a suitable, or safe, way to randomise samples.

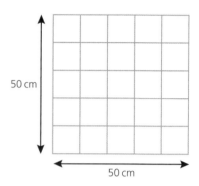

50 cm

50 cm

Figure 10.2 A gridded quadrat frame

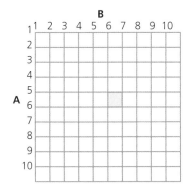

Figure 10.4 A grid of the sample area with coordinates

Now test yourself

TESTED ☐

6 Which statistical test should you use to test whether the difference between the mean number of organisms in two locations is significant?

Answer on p. 204

Randomising samples using a transect

You use this method if you suspect that environmental variables change within the sampling area, for example from the high tide mark to the low tide mark on a rocky shore.

1 Lay a tape measure across the sample area.

2a In a **belt transect**, you place a quadrat frame continuously along the tape measure and record the organisms within it.

2b In an **interrupted belt transect**, you place up to five quadrat frames to the same side of the tape at regular intervals, for example every 4 m along the transect (Figure 10.5).

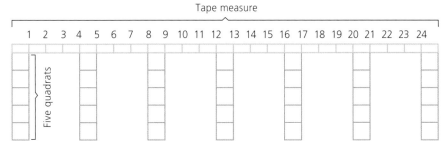

Figure 10.5 An interrupted belt transect

Now test yourself

TESTED ☐

7 Suggest how you could find the abundance of tadpoles in a pond.

Answer on p. 204

Method for assessing abundance of mobile organisms: the mark–release–recapture method

Animals that move quickly will not stay within a quadrat frame, so you need another method for estimating their abundance. You do this using the **mark–release–recapture method**.

1 Catch a sample of the animals from the sample area and count them (N_1).

2 Mark the animals in a way that will not harm them or make them more visible to predators.

3 Release the marked animals and allow sufficient time for them to disperse among the population from which they came.

4 Catch a second sample from the same population and count the number in the sample (N_2) and the number in the sample that are marked from the first sample (n).

This method assumes that the fraction of the total population (x) caught in the first sample, $\frac{N_1}{x}$, is the same as the fraction of marked individuals in the second sample, $\frac{n}{N_2}$, i.e.:

$$\frac{N_1}{x} = \frac{n}{N_2}$$

where x is the population size.

By transposing this formula, you can find the population size, x, as:

$$x = \frac{N_1 \times N_2}{n}$$

Now test yourself

TESTED

8 A student collected a sample of 12 snails from his garden. He marked them with quick-drying paint and released them back into his garden. After 3 days, he took a second sample of snails from the garden. Of the 15 snails, four carried his paint mark. Calculate the abundance of snails in this garden.

Answer on p. 204

This technique is based on the assumption that, during the period between samples:
● there are no births or deaths
● there is no migration
● the marking does not affect the behaviour of the released animals
● the released animals have time to mix randomly with the unmarked animals
● both samples are representative of the whole population

Energy transfer through ecosystems

Transfer of energy between trophic levels and to the environment

REVISED

As you saw in Topic 5, the light-dependent stage of photosynthesis is driven by light. Light itself consists of particles, called **photons**, that travel in waves.

When photons strike chlorophyll and boost a pair of electrons to a higher energy level, we can say that energy from the photons has been transduced to the electrons. We can extend this concept to the transfer of energy to ATP during chemiosmosis (page 99), from there to organic molecules synthesised by a plant and from there to any organism in higher trophic levels that consumes those organic molecules and uses their components to synthesise its own organic molecules.

Now test yourself

 TESTED ☐

9 What is chemiosmosis?

Answer on p. 204

A key principle in thermodynamics is that energy cannot be created or destroyed; it can only be transduced. This means that energy can only be 'used' once by an organism. The energy that is released by the hydrolysis of ATP might be 'used' to synthesise a polymer, for example. If more energy is released from ATP hydrolysis than is 'used' to synthesise the polymer, it is lost to the environment as heat. This idea is encapsulated in Figure 10.6.

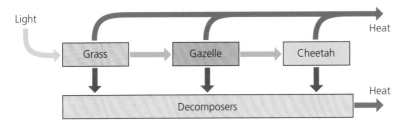

Figure 10.6 Energy flow through a simple food chain

Measuring productivity

Productivity is a measure of how much the biomass of a particular trophic level has increased:

- in a particular period of time
- for a particular area of an ecosystem

The increase in biomass can be expressed in one of two ways:

- dry mass
- the energy that would be released on complete combustion of that dry mass

You will use productivity in the latter context, i.e. the energy that would be released on combustion.

The productivity of producers is termed **primary productivity**.

During its life cycle, an annual plant will trap light and produce a large number of molecules via photosynthesis. The energy 'trapped' in these molecules represents the **gross primary productivity** during that year. The plant will use some of these molecules as respiratory substrates. They will not be there at the end of the year.

Alternatively, the plant will use some of these molecules to make new cells or to deposit as food stores. They will be there at the end of the year and form the biomass that will be eaten by primary consumers or used by decomposers following the death of the plants. The energy 'trapped' in these molecules represents the **net primary productivity**.

> **Primary productivity** is a measure of the light energy fixed by photosynthesis in a given area and in a given time.

> **Gross primary productivity** is the energy that would be released on complete combustion of the total biomass produced by the plants in a particular time in a particular area.

> **Net primary productivity** is the gross primary productivity minus respiratory losses.

We can express these ideas in a simple equation:

NPP = GPP – R

where **NPP** represents net primary productivity, **GPP** represents gross primary productivity and **R** represents respiratory losses.

Now test yourself   TESTED ☐

10 In what units would you represent net primary productivity in this equation?

Answer on p. 204

Figure 10.7 summarises data obtained from an ecosystem in the USA. The ecologists measured the energy values of light falling on each square metre of the ecosystem and the productivity of the organisms in each trophic level.

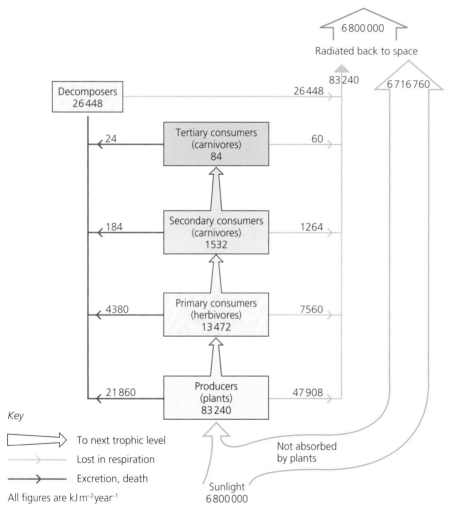

Figure 10.7 Estimates of gross and net productivity in the Silver Springs ecosystem (USA). The figures are given as kJ m⁻²year⁻¹

Now test yourself   TESTED ☐

11 Use information from Figure 10.7 to modify the equation for net primary productivity to represent net productivity for the primary consumer trophic level.
12 Figure 10.7 shows that 6 800 000 kJ m⁻²year⁻¹ fell on the ecosystem and 6 800 000 kJ m⁻²year⁻¹ were radiated from the ecosystem back into space. What important principle does this demonstrate?

Answers on p. 204

Calculating the efficiency of energy transfer

You can use data like those in Figure 10.7 to calculate the efficiency with which organisms use the energy available to them. For example, the scientists measured the energy within:

- the light falling on the ecosystem as $6\,800\,000\,\text{kJ}\,\text{m}^{-2}\,\text{year}^{-1}$
- the gross productivity of the producer trophic level as $83\,240\,\text{kJ}\,\text{m}^{-2}\,\text{year}^{-1}$.

Using these data, you can calculate the efficiency with which the producers 'used' the available light, by using the following formula:

$$\text{efficiency} = \frac{83\,240}{6\,800\,000} = 1.22\%$$

Now test yourself

TESTED

13 Write a word equation to show how you could calculate the efficiency of producers.

Answer on p. 204

Clearly the plants in this ecosystem were not efficient in their ability to utilise the light that fell on them. Figure 10.7 shows why — most of the incident light is not absorbed by the plants because:

- much of the light is the wrong wavelength to be absorbed by the photosynthetic pigments (look back to Figure 5.8, page 101)
- some light does not fall on chloroplasts or does not fall on the leaves at all
- some of the light simply heats up the leaves

We can use the data in Figure 10.7 to calculate the efficiency of energy transfer between any two trophic levels.

Now test yourself

TESTED

14 Use the data for gross productivity in Figure 10.7 to calculate the efficiency of energy transfer from producers to herbivores in this ecosystem.
15 Suggest *two* ways in which chicken farmers could increase the efficiency of energy transfer by their chickens. Are there ethical issues raised by your suggestions?

Answers on p. 204

Ecological pyramids

You can see in Figure 10.7 that the productivity decreases at each trophic level. By the final trophic level, the entire gross productivity is lost in respiration and to decomposers; there is none left to pass to another trophic level. This limits the number of trophic levels that can be supported in any ecosystem.

We can represent this idea using pyramids to represent an ecosystem at any given moment.

- A **pyramid of numbers** represents the total number of organisms in each trophic level.
- A **pyramid of biomass** represents the total dry biomass of organisms in each trophic level.
- A **pyramid of energy** represents the energy that would be released by complete combustion of all the organisms in each trophic level.

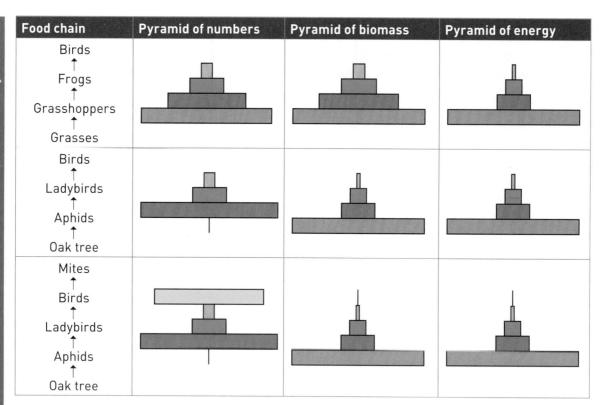

Figure 10.8 Pyramids of numbers, biomass and energy for three different ecosystems

Three examples of these pyramids are given in Figure 10.8. Notice that although pyramids of biomass and pyramids of energy always have the same shape, pyramids of numbers do not.

Although these pyramids provide easy-to-interpret information, they have their drawbacks:

- They all rely on effective sampling of the environment. For example, a tree is easy to count but the insects living on it are not.
- Dry biomass or the energy released by the complete combustion of dry biomass are better indicators of productivity but both involve killing organisms in the community.
- All measures represent a snapshot of communities at a particular time. Whilst, for example, a tree might be there all year, the insects and the birds that feed on it might not.

Now test yourself

TESTED ☐

16 If you were to add a bar representing decomposers to a pyramid of numbers in Figure 10.8, how wide would it be compared with the others? Explain your answer.

Answer on p. 204

Changes in ecosystems

The ecosystems that exist today did not always exist. They have developed over time by a process called **succession**.

- **Primary succession** occurs following colonisation of an environment that was originally organism-free, such as bare rock.
- **Secondary succession** occurs following recolonisation of a previous ecosystem that was destroyed, for example by fire.

Succession is a series of changes of the populations forming a community that occur as a result of changes in the abiotic and biotic factors of the ecosystem.

Now test yourself

TESTED ☐

17 Why do the colonisers of a habitat disappear from the community during succession?

Answer on p. 204

Figure 10.9 represents primary succession in the UK. Succession starts with bare rock — a pretty inhospitable substratun

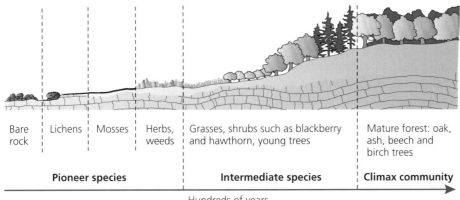

| Bare rock | Lichens | Mosses | Herbs, weeds | Grasses, shrubs such as blackberry and hawthorn, young trees | Mature forest: oak, ash, beech and birch trees |

Pioneer species · · · · · · **Intermediate species** · **Climax community**

Hundreds of years

Figure 10.9 Primary succession in the UK

1 Colonisation of this environment is possible only by **pioneer species** that can tolerate the harsh conditions. In this case, the pioneers are lichens.

2 Growth of the pioneer species, together with physical weathering, breaks up the rock. Death and decay of the lichens adds organic matter; a soil begins to develop.

3 Mosses can now colonise the developing soil. They out-compete the lichens, which disappear. Death and decay of the mosses adds more organic matter to the developing soil, allowing herbaceous plants to grow that, in turn, out-compete the mosses.

4 This cycle of new species entering the ecosystem, out-competing and displacing the existing species and causing further changes that make the environment less hostile continues, until trees are able to grow.

5 As more, and bigger, plants colonise the area, they provide more habitats and a greater variety of food sources for animals. The species richness increases and food webs become more complex.

6 Eventually, succession stops when a complex and stable **climax community** develops. The type of climax community that develops depends largely on the abiotic factors of the ecosystem. In much of the UK, the climax community is the deciduous forest represented in Figure 10.9. Other ecosystems in the UK include, grassland, moorland, pond, lake, rocky shore and estuary.

> A **climax community** is the stable group of populations at the end of ecological succession.

> **Typical mistake**
>
> Students often write that the more diverse the community the more food there is for animals. Examiners expect you to refer to more food types, rather than just more food.

Now test yourself

TESTED ☐

18 How would you expect the species of animals to change during the succession shown in Figure 10.9? Explain your answer.

19 Suggest how succession following destruction of a deciduous forest by fire would differ from the primary succession that led to the development of the forest.

Answers on p. 204

The effects of biotic and abiotic factors on population size

You saw in Topic 6 the typical growth curve for a bacterial population grown in batch culture (Figure 6.5, page 115). This growth curve would also apply to populations of other organisms in a natural habitat:

- The exponential phase would follow the introduction of a population into a new habitat.
- The stationary phase would be the maximum population that could be supported in that environment, the so-called **carrying capacity**.

The carrying capacity is determined by the effects of:
- **abiotic factors** — the chemical and physical features of the environment, such as pH, light intensity, temperature and **edaphic factors**
- **biotic factors** — the biological factors, such as intraspecific competition, interspecific competition, availability of food and the presence of predators and parasites

You learned in Topic 3 that each species can tolerate only a limited range of each abiotic and biotic factor, giving rise to the **ecological niche** of that species (Figure 3.7, page 58).

> The **carrying capacity** represents the maximum population size that can survive indefinitely in a given environment.
>
> **Abiotic factors** are the chemical and physical features of an environment.
>
> **Edaphic factors** are abiotic factors relating to soil, such as concentration of inorganic ions, water content and oxygen concentration.
>
> **Biotic factors** are those resulting from the presence of other organisms.

Human effects on ecosystems

The detrimental impact of humans on ecosystems is a direct result of the exponential growth of the human population. As our world population grows, we place greater demands on our environment for safe places to live, nutritious food, fuel with which to cook our food and manufactured items that we increasingly consider vital for a comfortable life. Table 10.3 summarises some of the detrimental effects that humans have on ecosystems around the world.

> **Typical mistake**
>
> Students sometimes try to hide their confusion between intraspecific and interspecific competition by writing a word with a spelling somewhere between the two. Examiners will expect you to spell the terms correctly and to show understanding of them.

Table 10.3 **Examples of how humans have a harmful effect on ecosystems**

Human effect on ecosystem	Examples
Destruction of entire ecosystem	• Clearing entire communities to provide land on which to build homes, work places, roads etc.
Removal of dominant species in climax community	• Clearing of trees in areas of forests to provide: 　– timber for building 　– arable land for crops 　– pastureland for grazing animals
Severe depletion, or elimination of a population within a community	• Overfishing of cod in the North Atlantic • Hunting to near extinction of the northern white rhino • Elimination of competitors from food chains, e.g. introduction of myxomatosis into Britain to kill rabbits
Prevention of ecological succession to produce a different (deflected) climax ecosystem	• Cultivation of monocultures • Burning of moorland heather to facilitate grouse shooting • Weeding of gardens
Pollution	• Aerial pollutants from burning fossil fuels • Water pollution from runoff of fertilisers and pesticides from agricultural land • Land pollution by dumping mine wastes in slag heaps • Pollution of land and water by non-biodegradable plastics
Introduction of non-indigenous species	• Repeated introductions of the grey squirrel in the UK, which have largely replaced the indigenous red squirrel • Introduction of plants, such as floating pennywort, into garden ponds, which have become important pests in waterways

Managing the conflict between human needs and conservation

REVISED

Since the middle of the twentieth century, groups of people have tried to influence policy makers to take action to ensure the sustainability of resources. Governments soon realised that international cooperation was needed to achieve the management of resources required to ensure their sustainability. This cooperation is ongoing, and has produced a number of international protocols and agreements.

- The Convention on International Trade in Endangered Species of wild fauna and flora (**CITES**) was set up in 1973. Its focus is the conservation, and prevention of trafficking, of endangered species. Among the species covered are elephants, whose tusks are removed to feed an illegal trade in ivory, and tigers, whose body parts are believed by some to have medicinal powers.
- In December 1995, a conference hosted by the **United Nations** adopted an agreement that individual states should cooperate to ensure conservation of stocks of fish by the optimum utilisation of fisheries resources. The contribution of the European Union has been to set strict fishing quotas in the seas around Europe.
- The Intergovernmental Panel on Climate Change (**IPCC**) focuses on measures to combat the effects of climate change. Among its successes were the **Kyoto Protocol**, signed in December 1997, and the outcome of the **Paris Climate Change Conference** in November 2015, in which countries made commitments to take action to limit global warming to below 2°C.

Now test yourself

TESTED

20 Name *two* gases that contribute to global warming and give *one* way in which human activities have contributed to an increase in the concentration of these gases in the atmosphere.

Answer on p. 204

Despite the number of countries attending IPCC summit meetings, you might believe from media coverage that global warming is a highly controversial issue. This is partly because the media tries to provide a balanced debate, inviting contributions from pro- and anti-global warming participants. In reality:

- the vast majority of members of the scientific community agree that global warming is real and is the result of human activities, especially the release of carbon dioxide and other greenhouse gases
- the vast majority of world politicians accept the opinion of the scientific community
- a small number of scientists, politicians, industrialists and journalists, often dubbed 'global warming deniers', disagree with the scientific community

Look back to Topic 3 to remind yourself how the scientific community validates evidence.

Exam practice

1 A scientist investigated the primary productivity of a field of maize.
 At the end of a growing season, he estimated that 1 hectare (10 000 m²) of the field of maize contained 25 000 maize plants. He dried several of these plants in an oven at 105°C until their mass became constant. From this, he estimated the mass of maize plants in the field to be 14 800 kg ha⁻¹.

(a) How could the scientist estimate the number of maize plants in the field? [3]

(b) Explain why he dried samples of the plants in an oven at 105°C until their mass became constant. [3]

 The scientist then found the rate of respiration of maize plants by measuring the rate at which they released carbon dioxide. He made this measurement during the night.

(c) Explain why he did this at night. [2]

 The scientist then converted his results into energy values. These are shown in the table.

Measured variable	Energy value/kJ ha⁻¹year⁻¹
Maize plants	26 000
Respiration of maize	8000
Sunlight falling on maize crop	2×10^6

(d) (i) Calculate the efficiency of gross primary productivity of this maize crop. Show your working. [2]

 (ii) Suggest *two* reasons for the low efficiency of gross primary productivity. [2]

2 (a) Explain the term ecological niche. [2]

 As their name suggests, flour beetles live in, and feed on, flour.

 The table shows the results of replicate experiments in which equal numbers of two species of flour beetle were kept in the same fixed mass of flour.

Temperature/°C	Relative humidity/%	Number of times each species died out	
		Tribolium castaneum	*Tribolium confusum*
34	70	0	10
34	30	9	1
29	70	2	8
29	30	8	2
24	70	7	3
24	30	10	0

(b) What was the total number of replicate experiments carried out? [1]

(c) Why was it important that each replicate began with a fixed mass of flour? [2]

(d) What can you conclude from the results in the table? [4]

3 (a) Distinguish between the terms:
  (i) population and community
  (ii) ecosystem and habitat [2]

(b) Put a tick against each of the features in the table to show the change that will occur during ecological succession from bare rock. [5]

Feature	Stays the same	Decreases	Increases
Biomass			
Nutrient content of soil			
Primary productivity			
Rate at which species replace each other			
Species diversity			

4 A number of international organisations take a leading role in balancing human needs and conservation. The work of each international organisation is aided by scientific advisers.

(a) The CITES treaty is one of the oldest agreements, supported by 180 member countries. Outline its role. [3]

(b) How do scientists ensure that the advice they provide is valid? [4]

Answers and quick quiz 10 online

ONLINE

Summary

The nature of ecosystems

- An ecosystem comprises the chemical and physical surroundings and the community that lives within them.
- Populations within these communities can be classed into different trophic levels — producers, consumers and decomposers. Producers use photosynthesis to increase their dry mass. This is called primary production and provides food for consumers and decomposers.
- Decomposers are microorganisms. Their hydrolytic activity results in nutrient cycles.
- Trophic levels can be represented by pyramids of numbers, dry mass (biomass) or energy content that all provide a snapshot of the community at a particular time.
- Methods for investigating the distribution and abundance of populations within an ecosystem include:
  - using samples, taken with, for example, randomly placed quadrats or quadrats along a transect
  - counting the number of organisms, estimating their percentage cover or using ACFOR scales
  - the mark–release–recapture method for mobile animals

Energy transfer through ecosystems

- Gross primary productivity (GPP) is the rate at which primary production occurs. The productivity once losses due to respiration (R) are taken into account is the net primary productivity (NPP), i.e. GPP = NPP + R.
- The efficiency of energy transfer between trophic levels can be calculated.
- The efficiency of photosynthesis is generally low, with very little of the incident sunlight being absorbed by plants.
- The energy efficiency of consumers is dependent on the metabolic requirements of the animals in each population and is lower in endotherms than in ectotherms.
- The energy trapped by a community is eventually lost to the environment as heat. These energy losses limit the number of trophic levels that can be supported in any ecosystem.

Changes in ecosystems

- Ecosystems change over time, a process called ecological succession.
- Bare habitats are initially inhabited by populations of colonisers. The actions of these colonisers change the nature of the environment, making it less hostile to other populations that become established and further change the environment.
- As succession proceeds, the earlier populations disappear from the community as they are replaced by populations that are better competitors.
- Each population within a community is limited to a sustainable size by a combination of abiotic and biotic factors.
- The process of succession is often deflected by human activities, including agriculture.

Human effects on ecosystems

- The growth of the human population has been sustained by exploiting natural ecosystems. This exploitation has had detrimental effects on many ecosystems, including overfishing.
- International agreements and treaties have been introduced to ensure the sustainability of resources by the effective management of human needs and conservation of natural ecosystems.
- The scientific community contributes to these agreements and treaties by providing information based on valid, peer-reviewed evidence.

Now test yourself answers

Chapter 1

1 $C_{12}H_{22}O_{11}$

2 It would create a water potential gradient into the plant cells.

3 The polysaccharides are hydrolysed at their ends. The more ends, the more enzymes can attach and hydrolyse the polysaccharides.

4 A polymer is a chain of repeated molecules of the same type (monomers). The components of a triglyceride are not the same type of molecule.

5 Three — primary (order of amino acids), secondary (each polypeptide is a helix) and quaternary (four polypeptides in the molecule).

6 Deoxyribose has a hydrogen atom in place of a hydroxyl group on carbon-2.

7 It is a covalent bond between a phosphate group and the carbon-3 of one nucleotide and the carbon-5 of the next nucleotide in the chain.

8 (a) The tertiary structure of the active site of DNA polymerase is complementary only to the carbon-3 end of a nucleotide.

   (b) As the two strands in a DNA molecule are antiparallel, they will develop in opposite directions (and one will grow continuously but the other will grow discontinuously).

9 It has the base thymine (RNA would have uracil).

10 It is AUG. (As mRNA is produced from the antisense strand of DNA, its base sequence is the same as that of the sense strand of DNA, except that RNA contains uracil in place of thymine.)

11 An exon is coding whereas an intron is not.

12 Prokaryotic DNA lacks introns/is all coding.

13 GAU and CGG

14 ATG. It is the only base triplet that codes for methionine (see Table 1.2).

15 Sucrose and lactose molecules are not complementary to the tertiary structure of the active site of maltase.

16 All the substrate has been converted to product.

17 2.5 arbitrary units per minute

18 The curve should have the converse shape of the current curve. It will start at a high level of substrate and fall to zero.

19 An anion

Chapter 2

1 The nucleus contains the genes that encode proteins. Ribosomes on the rough endoplasmic reticulum produce proteins. The Golgi apparatus modifies proteins and produces vesicles that carry the proteins to the cell surface membrane for secretion.

2 The toxin will kill or deter the herbivore that caused the cell damage.

3 The (phospholipid bilayer) cell surface membrane and 70S ribosomes are found in both types of cell.

4 The magnification is ×12500 (25 mm = 25000 µm; and 25000/2 = 12500).

5 Antibiotics work by interfering with the metabolism of the infective bacterium. Viruses lack a metabolism.

6 Treatment, such as with antibiotics, is ineffective on viruses, so preventing the spread of the viral disease is the most effective way to stop other people contracting it.

7 Latency. The virus becomes part of the DNA in the cells of the sufferer and, when stimulated by some form of stress, becomes detached and takes over the host cells' metabolism at a later time.

8 (a) 2^{23}

   (b) 2^{46} (i.e. $2^{23} \times 2^{23}$)

9 The most likely suggestion is that cells normally survive only if they have both members of each pair of homologous chromosomes. The sex chromosomes are not homologous. The Y chromosome carries very few genes, yet a male survives with only one X chromosome.

10 It has a large volume of cytoplasm, containing the cell organelles and fuel stores that the zygote will need for growth.

11 7 ($1 \rightarrow 2 \rightarrow 4 \rightarrow 8 \rightarrow 16 \rightarrow 32 \rightarrow 64 \rightarrow 128$ cells)

12 It increases the surface area for the exchange of substances between the embryo/fetus and mother's blood supply. It anchors the blastocyst firmly within the uterus lining.

13 Mammalian gametes are cells, but gametes of the flowering plant are nuclei.

14 Pollination occurs when a pollen grain lands on the stigma of a suitable plant and begins to grow. Fertilisation is the fusion of one male

nucleus with the female nucleus and the other male nucleus with both polar nuclei.

15 (a) It organises the activities/growth of the pollen tube.

(b) The pollen tube digests some of the tissues of the stigma/style, absorbs the products of digestion and uses them to manufacture new molecules that enable it to grow.

16 They both contain seeds — containing the embryo plants — enclosed within a modified ovary wall.

Chapter 3

1 No one uses Latin, so it is not likely to be corrupted by colloquialisms.

2 They are different species. In fact, they belong to different genera (plural of genus). The common name of 'robin', used in both the UK and the USA, is not helpful to taxonomists.

3 It is not possible to test the ability to interbreed when: considering extinct organisms/fossils; considering dead/preserved specimens; only one organism is available; all the available organisms are of the same sex; dealing with organisms that reproduce asexually.

4 It is a natural system. Biologists are trying to establish evolutionary relationships between organisms.

5 DNA polymerase catalyses the formation of (phosphodiester) bonds between adjacent DNA nucleotides.

6 DNA polymerase catalyses the formation of a phosphodiester bond between the hydroxyl group on carbon-3 of the terminal nucleotide and the phosphate group on carbon-5 of the nucleotide to be added.

7 A DNA molecule is too large, especially in eukaryotic cells.

8 The phosphate group (PO_4^{3-}) in each nucleotide

9 It tests a prediction made on the basis that a hypothesis is true. If the results of the investigation are consistent with the prediction and can be replicated, they lend support to that theory. The possibility always exists, however, that the test of another prediction from the same hypothesis might produce results that do not support the hypothesis. Consequently, although scientists strongly believe in many hypotheses, they must always be prepared to accept new results that show they were wrong.

10 The short answer is to ensure that the investigators did not bias the investigation or conclusion. The more detailed answer is to ensure the investigators made a valid prediction from their hypothesis, designed a valid test of that prediction, carried out a

reliable procedure, analysed their results with the use of appropriate statistical tests and so ensured that their conclusions are valid. If not, the paper should not be accepted for publication in a scientific journal that is trusted by other scientists.

11 Either to overcome the activation energy of the reaction or because condensation reactions are energy-consuming.

12 They comprise the same type and number of atoms but these are arranged in a different way.

13 Four of the kingdoms (Protoctista, Fungi, Plantae and Animalia) would all be placed in the domain Eukarya. (Only the classification of the kingdom Prokaryotae would need to be changed as this comprises the domains Archea and Bacteria.)

14 Mutation

15 The organism with the selective advantage: survives (to reproductive age) whereas the other does not; uses less of its energy in maintenance/has more energy for reproduction than the organism without the selective advantage.

16 In a population that is well adapted to its niche, stabilising selection will act on any deviation from the norm. A random mutation is likely to make a bacterium less well adapted to its current niche. Only in the presence of antibiotic X is the mutation likely to confer a selective advantage.

17 Diversity in the gene pool of a large population would be greater than in a small population. In the latter, reproduction between few individuals results in most organisms having the same alleles of each gene. You will learn more about this if you follow the second year of an A-level biology course.

Chapter 4

1 Fatty acids

2 The thicker the exchange surface, the greater the resistance to the movement of particles. The greater the number of ions/molecules on one side, the more frequent their collisions with, and movement through, the membrane.

3 The cell will gain water. Water moves by osmosis from a higher water potential (–5 kPa) to a lower water potential (–10 kPa).

4 The ATP is hydrolysed as fast as it is made.

5 The rate of respiration must be less than the rate of photosynthesis.

6 Being microscopically small, its surface area to volume ratio is high.

7 It reduces water loss by evaporation from tissues.

8 The length of the diffusion pathway of a very large insect would severely reduce the efficiency of gas exchange.

9 As the blood flows in one direction, it is always surrounded by water with a higher concentration of oxygen and lower concentration of carbon dioxide, so both concentration gradients remain steep right across the gill lamellae.

10 Without the support of water they would collapse, severely reducing their surface area to volume ratio.

11 It reduces the distance over which gas exchange must occur.

12 A suitable flow chart would be:

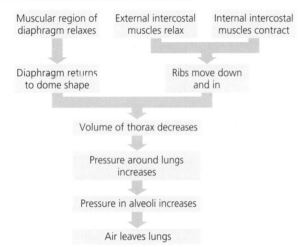

13 When blowing air out of the mouth.

14 (a) Friction against the walls of the blood vessels.

(b) These vessels have a combined surface area that is much greater than the larger vessels, so the effect of friction is greater.

15 The heart is located in the thorax. The pressure in the thorax enabling the prolonged exhalation might be greater than the blood pressure pushing blood back to the heart.

16 It allows the ventricles to fully empty before ventricular systole begins.

17 It prevents the blood clotting in the containers, as calcium ions are needed to activate thrombin.

18 It would make the plasma more viscous, slowing blood flow if haemoglobin were free in the plasma.

19 Active muscles respire rapidly, producing carbon dioxide. The increased carbon dioxide concentration results in oxyhaemoglobin dissociating more readily. As a result, muscle cells that are respiring most rapidly gain more oxygen than cells that are less active.

20 Haemoglobin of the fetus will gain oxygen from the maternal haemoglobin across the placenta. Muscle cells will 'store' oxygen in the form of oxygenated myoglobin.

21 Loss of water from the plasma and friction against the walls of the capillaries.

22 Their concentration of blood proteins would be less, so the oncotic pressure in the capillaries would be less. At the arteriole end, this would result in a greater net outward pressure, so more tissue fluid would form. At the venule end, the net inward pressure would be less, so less tissue fluid would be reabsorbed.

23 Tonoplast of first cell → cytoplasm → cell sirface membrane → cell wall → cell wall of second cell → cell sirface membrane → cytoplasm → tonoplast of second cell

24 Lignin helps the walls withstand the tension of the water within them.

25 The plants will be photosynthesising, so their stomata will open more often.

26 Starch grains will not affect the water potential of the mesophyll cells, so the influx of water that provides the pressure for translocation will be less or will not occur. This does not support the mass-flow hypothesis.

Chapter 5

1 Hexose molecules are too large to pass through the mitochondrial membrane *or* there are no glucose transporter proteins in the outer mitochondrial membrane.

2 (a) They increase the surface area, allowing more of the molecules that form ATP.

(b) Like all reactions, the reactions of the Krebs cycle occur more rapidly in solution.

3 Phosphorylation makes the molecule more reactive.

4 Four: two from glycolysis and one from each of the two pyruvate molecules involved in the link reaction.

5 Being ions, they cannot diffuse through the phospholipid bilayer. The ATP synthase provides a specific protein channel through which they can pass by facilitated diffusion.

6 Figure 5.2 shows that glycolysis uses two molecules of ATP to phosphorylate one molecule of hexose but that four molecules of ATP are later produced. No further molecules of ATP

are formed in Figure 5.6, leaving a total of two molecules of ATP per molecule of hexose.

7 Like all membranes, the thylakoids are phospholipid bilayers, so lipid-soluble pigments will dissolve in them.

8 Having more than one pigment enables plants to absorb light of different wavelengths, so more light is absorbed.

9 The wavelengths of light falling on shaded plants are different from those falling on unshaded plants. For example, light falling on plants growing on the floor of a woodland will have lost the wavelengths absorbed by the leaves of the trees above them. The combination of pigments adapts the plants to the wavelengths of light falling on them.

10 In respiration, oxygen is the final electron acceptor in the electron transport chain. In the light-dependent stage of photosynthesis, oxygen is produced by photolysis of water.

11 The one labelled 2

12 Oxygen and reduced NADP

13 Emerson concluded that photosynthesis depends on two light-harvesting systems that absorb different wavelengths of light. You have learned that these are PSI and PSII.

14 The light-independent stage uses ATP and reduced NADP from the light-dependent stage; in the dark these will soon be used up and cannot be replenished.

15 Because the rate of photosynthesis increases when the light intensity is increased.

16 It increases the temperature and the carbon dioxide concentration.

17 The water, inorganic ion and pigment content of the photosynthesising cells.

Chapter 6

1 5, 6, 7, 8 and 9

2 Then heat would cause 'spitting' of droplets of liquid medium, contaminating the surroundings.

3 (a) Continuous, so that the useful product can be harvested continuously.

  (b) Batch, because only a small sample is required for storage and the refrigerator will slow the growth of the bacteria.

4 Your arrow should point to the end of the zigzag line.

5 You would flame the inoculating loop after each use and you would lift the lid of the Petri dish as little as possible.

6 1×10^{-3} (standard form is always $A \times 10^n$, where A is a number above 1 and below 10 and n is the number of places the decimal point moves).

7 Eight cells

8 $4 \times 10^{-3} \, mm^3$

9 Volume of inoculum and dilution factor of inoculum

10 Gram-negative bacteria have thinner walls of peptidoglycan than do Gram-positive bacteria. They also have an additional lipid and polysaccharide outer layer that Gram-positive lack. (Note that this question tested recall and understanding (assessment objective 1) of content from Topic 2.

11 Since they hang freely from the neck, they could transfer bacteria from one patient to another.

12 You could answer in one of two ways. Firstly, many people begin to feel well again and stop taking the antibiotic; sub-lethal doses of antibiotic provide a selection pressure favouring any chance mutation conferring resistance against that antibiotic. Secondly, many of the illnesses for which patients request antibiotic are caused by viruses and antibiotics do not affect viruses.

13 The hyphae of the stem rust fungi break the protective outer layer of the host plant, which will allow access to other pathogens.

14 4×10^8

15 Antigenic variability means that the parasite frequently changes the molecules on its surface that stimulate antibody formation. As antibodies are specific, the existing vaccine would not be complementary to the parasite's new surface molecules.

16 Digestion by an antigen-presenting cell is likely to result in several different peptides from a single pathogen. All these peptides will then be displayed on the surface membranes of antigen-presenting cells and each will be complementary to the receptors of different lymphocytes.

17 You might expect them to have: large numbers of mitochondria; well developed rough endoplasmic reticulum/large numbers of ribosomes; well developed Golgi apparatus.

18 When someone is in danger of dying before their own immune system is able to remove the pathogen or toxin. Correct examples would include following a venomous snake bite (toxin) or following a cut by a soil-contaminated garden tool (contaminated by the bacterium that causes tetanus).

19 T cells do not respond to antigens in solution but B cells do and so respond to antigens in the blood. Being proteins, antibodies cannot cross cell sirface membranes to get to the pathogens inside infected cells but T killer cells can rupture the cell surface membranes of infected cells, exposing the pathogens.

20 Correct answers will include groups of people with: undeveloped immune systems, such as babies and toddlers; weakened immune systems,

such as the elderly or people suffering protein deficiency; compromised immune systems, including HIV patients and cancer patients.

Chapter 7

1 Nucleus, mitochondria and chloroplasts

2 A very high temperature/95°C is used to separate the two polynucleotide chains in DNA and replication occurs at a high temperature/65°C; the DNA polymerase is present in the mixture/is re-used in each cycle, so it must not be denatured by these high temperatures.

3 Hydrolysis of the nucleoside triphosphate releases energy, so enabling formation of the phosphodiester bond.

4 The one at the bottom of the diagram/of the X-ray film, since this is the one that moved furthest from the cathode through the agarose gel.

5 The phosphate group/(PO_4^{3-}) — an anion

6 (a) ACGCATGTTC

   (b) TGCGTACAAG

7 A dideoxynucleotide containing adenine

8 Split the sequence into triplets and use a table like Table 1.2 (page 16) to see which amino acid is encoded by each triplet. The triplet ATG is always the 'start' triplet.

9 White blood cells (red blood cells lack a nucleus).

10 The direction in which genes are transcribed is termed 'downstream'. 'Upstream' refers to the region in the opposite direction from that in which the DNA will be transcribed.

11 Part of the DNA-transcription factor complex has a shape that is complementary to that of part of the RNA polymerase molecule.

12 The link reaction in aerobic respiration (Figure 5.3)

13 (a) The histone around which the DNA is wrapped

    (b) Cytosine of a nucleotide that is in a CpG sequence

14 tRNA carries specific amino acids to the ribosomes during the translation stage of protein synthesis.

15 Differentiation results as cells 'switch off' some of their genes, often as a result of epigenetic modifications as a result of epigenetic modifications and local 'cues' from neighbouring cells and tissues.

16 They possess 'self-antigens', so would not cause an immune response.

17 If proto-oncogenes are activated, they become oncogenes that stimulate rapid cell division. Rapid cell division produces a tumour that might damage other organs as it presses against them, or might develop into a cancer.

18 It contains DNA from two different organisms.

19 A sequence that is the same when 'read' in both directions of the antiparallel strands in a DNA molecule.

20 The same restriction enzyme will produce complementary unpaired bases (i.e. complementary 'sticky ends').

21 They have been able to grow on agar containing ampicillin, so they must have taken up the plasmid containing the gene for this antibiotic resistance.

22 Unlike cells from colonies 2 and 4, they were able to grow on agar containing tetracycline, so their gene for resistance to this antibiotic must still be intact.

23 Look for a change in the physiology or behaviour of the 'knockout' mice.

24 Knock out the gene that controls the healthy condition, thus inducing symptoms of the disease. Then administer the test therapy to see if it removes the symptoms of the disease.

Chapter 8

1 Random fertilisation of gametes

2 Gene mutation. (You might have answered horizontal gene transfer, which is correct, but is not subject content for this specification.)

3 0.25 (remember that probabilities have decimal values between 0.0 and 1.0).

4 The woman would have to be a carrier and her partner a haemophiliac.

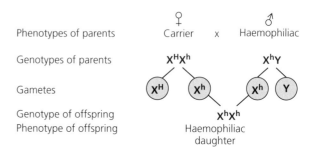

5 (a) Individual 8 is represented by a square so must be a male with genotype XY. The colour coding shows he is haemophilic so his X chromosome must carry the recessive 'h' allele, hence X^h

   (b) X^HX^h and X^HY

   (c) X^HX^h (her partner is not haemophiliac but she has a haemophiliac son).

6 The locus of that gene

7 It would be difficult to distinguish between the handwritten upper case and lower case of each letter (**C** and **c**; **W** and **w**) and could lead to you making an error in the exam or to the examiner being able to tell your upper case from your lower case letters.

8 There has been random assortment of these chromosomes during meiosis; all the gametes have been involved in random fertilisation; there is a large number of offspring (reducing the influence of chance events on a small sample).

9 The greater the distance between the two genes, the greater the frequency of crossing over.

10 The ratio of phenotypes would depend on the number of gametes affected by crossing over and you cannot know in advance how many crossovers will occur.

11 Genes with loci that are not on the same chromosome.

12 Divided the 320 offspring from the actual cross into a 9:3:3:1 ratio.

13 Because the total number of offspring is 320, so when we know the values of three of the categories, the fourth value is fixed.

14 Within a single population, organisms with a particular phenotype have more offspring than others with a different phenotype.

15 Organisms with the two phenotypes might become two reproductively isolated populations on which natural selection has different effects. The two populations become separate species when they fail to interbreed and produce fertile offspring.

16 Chance events are less likely to affect the expected outcome in a large sample than in a small sample.

17 A genetic bottleneck caused low genetic diversity in cheetah populations. The genetic diversity is so low that the self-antigens of the donor cheetah are not recognised by the recipient cheetah as foreign.

18 The islands were populated from the mainland by a small number of mice (or perhaps even a single pregnant female) that probably survived on driftwood. The coloniser/s had only a small sample of the gene pool of the mainland population. This is an example of the founder effect.

19 Valid suggestions include: the population is large; there is no movement of organisms into the population (immigration) or out of the population (emigration); the individuals reproduce sexually and mating is random; an individual's genotype does not affect its chances of breeding successfully; gene mutation does not occur.

20 Every member of that population has two alleles of that gene, so the frequency of these alleles must be 100% or, as a decimal value, 1.0.

21 (a) 0.16

   (b) Frequency of the recessive allele is $\sqrt{0.16} = 0.4$; frequency of the dominant allele is $1 - 0.4 = 0.6$.

   (c) The frequency of heterozygotes is $2 \times 0.6 \times 0.4 = 0.48$.

22 Find the allele frequency of one or more genes over a period of time. If the allele frequency shows a significant change, the population might be evolving.

Chapter 9

1 Changes in blood glucose concentration would affect any one of: blood viscosity and, hence, blood flow; blood water potential and, hence, osmosis into or out of cells; availability of respiratory substrate and, hence, regeneration of ATP within cells.

2 The cytokines activate more T helper cells, which, in turn, secrete yet more cytokines (page 121).

3 Condensation

4 Liver cells (hepatocytes)

5 Since steroids are lipids, they can dissolve in the phospholipid bilayer of cell surface membranes and diffuse through the membrane.

6 A protein that attaches to the promoter of a gene, enabling RNA polymerase to attach to the gene and begin transcription.

7 (a) Auxins and gibberellins work together in promoting cell elongation in root and in suppressing the growth of lateral buds.

   (b) Auxins and cytokinins have opposite effects on the growth of lateral buds.

8 Auxins

9 Because it is the short period of darkness that is the key to flowering in these plants. They are stimulated to flower by P_{FR}. During a short night P_{FR} remains in the plant and flowering occurs. A brief flash of light during a long night is sufficient to prevent the loss of P_{FR}, allowing flowering.

10 The myelin (a lipid) around the axons

11 The dendron carries impulses towards the cell body whereas the axon carries impulses away from the cell body/to the target organ.

12 If ion exchanges fail to result in a membrane potential of $-55\,mV$, no action potential occurs. If the threshold of $-55\,mV$ is reached, a full action potential occurs.

13 Positive feedback, because it causes yet more voltage-gated sodium ion channels to open.

14 Electromagnetic attraction

15 The membrane immediately behind the most recent action potential (shown in grey in Figure 9.10) is still in its refractory period (Figure 9.9), so only the membrane in front of the action potential in Figure 9.10 can be depolarised.

16 The ability to see fine detail.

17 An image 'in the corner of your eye' falls on the periphery of your retina. This is where rods are most dense and, since rhodopsin is degraded

by dim light, you see the star. When you look directly at the star, its image falls on the fovea of your retina. Here you have only cones. The pigment in cones is not degraded by dim light. So you do not 'see' the star.

18 (a) At the top of the spinal cord

(b) In the wall of the right atrium

19 Condensation

20 When someone has a diet with low carbohydrate content but high protein content (e.g. the Atkins diet).

21 Your answer should include water, glucose, amino acids, oxygen, carbon dioxide and inorganic ions, but exclude proteins.

22 They reduce the resistance to the flow of ultrafiltrate.

Chapter 10

1 (a) A population comprises only one species, whereas a community comprises many species/many populations.

(b) An ecosystem includes the community and its surroundings, whereas a habitat describes only the type of environment in which organisms are found.

2 The dry mass represents the products of metabolism/growth, whereas the mass of water does not. Alternatively, the mass of water in an organism can vary, even during a single day, whereas the dry mass will not show such variation.

3 Primary consumers

4 Bacteria and fungi

5 (a) Percentage cover — you cannot see individual lichens.

(b) Individual counts — they are large enough to count and do not move.

6 Student's *t*-test

7 Produce a gridded map of the pond; use a random number generator to find coordinates of the grid; take a fixed volume of water from the parts of the pond within the generated

coordinates; count the number of tadpoles in each sample; calculate the mean number of tadpoles per fixed volume of water.

8 45 snails

9 The production of ATP associated with the diffusion of protons (H^+ ions) through ATP synthase molecules embedded in the inner membranes of chloroplasts and mitochondria.

10 Your answer should include units for energy, area and time interval, for example $kJ\,m^{-2}\,year^{-1}$.

11 Net secondary productivity = gross secondary productivity − (respiration + faeces + urine)

12 Although energy can be transduced, i.e. changed from one form to another, it cannot be created or destroyed, which is the first law of thermodynamics.

13 Your word equation would include these concepts:

$$\frac{\text{gross primary productivity}}{\text{incident light energy}} \times 100\%$$

14 $\frac{13\,472}{83\,240} \times 100\% = 16.2\%$

15 They could keep the chickens indoors, reducing heat loss from the chickens. They could reduce the movement of the chickens, reducing energy involved in muscle contraction. Keeping chickens in restricted indoor conditions raises ethical issues about the quality of life of the chickens as well as increasing the risk of transmission of pathogens between chickens.

16 The bar would be much broader than the others, since the decomposers are microorganisms.

17 The organisms in the later populations are better competitors.

18 They would increase in number as the number of food sources increased and the number of resting places/nesting places increased.

19 Soil is already present so plants, rather than lichens, can be the pioneer species. Some heat-resistant seeds will have survived the fire and will germinate to colonise as pioneers.

20 Carbon dioxide — burning of fossil fuels; methane — paddy fields/intestines of cattle.